CYNGOR CEFN GWLAD CYMRU

COUNTRYSIDE COUNCIL FOR WALES

GWASANAETH LLYFRGELL LIBRARY SERVICES

**Rhanbarth de Cymru
South Wales Region**

Countryside Council for Wales
RVB House
Llys Felin Newydd
Phoenix Way
Swansea Enterprise Park
Llansamlet
SWANSEA
SA7 9FG

The Natural History of
BADGERS

Christopher Helm Mammal Series
Edited by Dr Ernest Neal, MBE, former President of the Mammal Society

Already published:

The Natural History of Antelopes
C.A. Spinage

The Natural History of Deer
Rory Putman

The Natural History of Otters
Paul Chanin

The Natural History of Seals
W. Nigel Bonner

The Natural History of Squirrels
John Gurnell

The Natural History of Weasels and Stoats
Carolyn King

The Natural History of Whales and Dolphins
P.G.H. Evans

Forthcoming titles include:

The Natural History of Hibernating Bats
Roger Ransome

The Natural History of Moles
Martyn L. Gorman and David Stone

The Natural History of Shrews
Sara Churchfield

The Natural History of Wild Cats
Andrew Kitchener

The Natural History of
BADGERS

Foreword by Sir David Attenborough FRS

ERNEST NEAL

CHRISTOPHER HELM
London

© 1986 Ernest Neal
Reprinted 1987, 1990

Line illustrations by Michael Clark

Christopher Helm (Publishers) Ltd, Imperial House,
21–25 North Street, Bromley, Kent BR1 1SD

ISBN: 0–7470–2412–X

A CIP catalogue record for this book is available from the British Library

All rights reserved. No reproduction, copy or transmission of this publication may be made without written permission.

No paragraph of this publication may be reproduced, copied or transmitted save with written permission or in accordance with the provisions of the Copyright Act 1956 (as amended), or under the terms of any licence permitting limited copying issued by the Copyright Licensing Agency, 7 Ridgmount Street, London WC1E 7AE.

Any person who does any unauthorised act in relation to this publication may be liable to criminal prosecution and civil claims for damages.

Printed and bound in Great Britain

Contents

List of colour plates	vi
List of figures	vii
List of tables	x
Acknowledgements	xi
Series editor's foreword	xiii
Foreword by Sir David Attenborough FRS	xv
Preface	xvii
1. Badgers: studying them in the field	1
2. General characteristics	15
3. The badger's home and environs	39
4. Distribution and choice of habitat	65
5. Sett tenants and associations with other animals	79
6. The special senses	87
7. Activity patterns	95
8. Food and feeding behaviour	109
9. Social life	139
10. Reproduction and development	159
11. Numbers	181
12. Badgers and man	195
13. The American badger	209
14. Badgers of Asia and Africa	215
Appendix I Skeletal criteria for determining the age of badgers. Modified from Hancox (1973)	224
Appendix II Road underpasses for badgers	225
Appendix III Badgers and the law in Britain	226
Bibliography	229
Index	236

Colour plates

1 Sow with two well-grown cubs; July
2 Badgers playing; May
3 Boar, sow and cub relaxing after emergence; July
4 Adult erythristic sow
5 Adult albino boar
6 Grooming session
7 In red sandstone districts, the coats of badgers soon become stained by the soil, through constant digging
8 Sow bringing back a bundle of grass bedding
9 Badger in day nest; June
10 Boar setting scent on back of sow
11 Woodland sett showing the ground outside worn hard and smooth by cub play. Mud marks may be seen on the scratching tree
12 Badger nest in a hollow tree with bedding overflowing; Sweden, August
13 An 'up-and-over' made by badgers when crossing a hedgebank bordering a Somerset lane; March
14 Remains of a wasps' nest dug out the previous night; September
15 Sow with young cubs; May
16 Cub aged about eight weeks, at sett entrance; April
17 Sow suckling her three cubs
18 American badger, *Taxidea taxus*; North America
19 Hog badger, *Arctonyx collaris*; South-East Asia
20 Honey badger, *Mellivora capensis*; north Kenya

Figures

1.1	Badgers of the world	2
1.2	World distribution of badgers	3
1.3	Radio tracking of badgers. a) Badger fitted with transmitter; b) harness and transmitter	11
1.4	Apparatus for the location and observation of badgers wearing transmitters. a) Aerials, receiver and earphones; b) viewer, power pack and binocular viewer	12
1.5	Watching badgers at night using infra-red binoculars	13
2.1	Head of an old boar showing rhinarium and whisker pattern	16
2.2	a) Fore foot; b) hind foot. Photographs of the same badger	17
2.3	Footprint in soft mud, showing heel mark	18
2.4	Badger climbing an elder after slugs	20
2.5	Semi-albino	28
2.6	Adult sow	29
2.7	Bacula. Four examples of known age	30
2.8	Skeleton of a badger	31
2.9	Adult boar badger skull; lateral and anterior views	32
2.10	Side view of skull to show direction of main fibres of the temporalis muscle, which gives great leverage to the jaw	33
2.11	Development of the skull up to one year, showing formation of sagittal crest	34
2.12	Dentition of cub aged 15 weeks, showing permanent teeth all present with extra canine and premolar of the milk dentition still not shed	36
2.13	Tooth wear; stages in exposure of dentine	37
3.1	Badger sett in elder thicket, showing spoil heap	40
3.2	Plan of a large sett in Gloucestershire, covering area of 35×15 m	41
3.3	Digging	45
3.4	Bedding gathering	47
3.5	Number of nights per month when digging was known to have taken place at one permanently occupied sett near Taunton, Somerset	50

LIST OF FIGURES

3.6	Number of nights per month when bedding collecting was known to have taken place at the same sett as for Figure 3.5	51
3.7	The relationship between bedding collecting and digging activity, incorporating the same data as for Figures 3.5 and 3.6	52
3.8	Bedding being aired in winter	55
3.9	Badger asleep in a day nest	56
3.10	Leaving hair on barbed wire above a path	57
3.11	A deciduous wood in the Quantock Hills, Somerset, containing four small setts belonging to the same social group	58
3.12	Badgers often clamber over a dead tree trunk if it straddles a main path	59
3.13	Badger path from wood into pasture; March	60
3.14	When badgers clamber over stone walls they regularly use the same place, wearing away the lichens in the process	61
3.15	Scratching an elder tree by the sett	63
4.1	Distribution of setts in relation to soils and landscape (Cotswold Scarp)	71
4.2	Sett dug in soft stratum under hard, impervious slates	72
6.1	Brain of badger; dorsal view	87
6.2	Badgers scent the air cautiously when emerging from the sett	92
7.1	Times (GMT) of first emergences when undisturbed. a) Throughout the year; b) during winter	96
7.2	Returning after a forage in the snow	101
7.3	Cautious emergence at dusk; three adults and a 12-week-old cub	102
7.4	Times of return to sett in the early morning	105
8.1	Cubs foraging below a flowering elder	111
8.2	Badgers on their way to their feeding grounds	112
8.3	Frequency of occurrence (per cent) of main food categories throughout the year in 3,846 dung samples	113
8.4	Diet of badgers in lowland Britain in the wet summer of 1963	115
8.5	Diet of badgers in lowland Britain in the very dry summer of 1964, for comparison with Figure 8.4	116
8.6	The stomach contents of a single badger . . . over 200 earthworms	118
8.7	Beetle larvae are commonly dug up and eaten	120
8.8	The gape shows how the nose projects beyond the lower jaw	121
8.9	Badger bringing back food to the sett	125
8.10	Foraging in an oat field	133
8.11	When badgers knock down oats, the straws usually lie in a criss-cross manner	134
8.12	Badger drinking from a hollow in a sycamore	137
8.13	Drinking	138
9.1	Frightened cub running back to sett, hair raised	140
9.2	Sub-caudal region of adult badgers	143
9.3	Scent marking at dung pits	145
9.4	Badger family	146
9.5	Sow emerging	147

LIST OF FIGURES

9.6	Territories of badgers of Wytham, as shown by recoveries in dung pits of coloured food markers presented on the setts (1–16)	149
9.7	Dung pit area at territorial boundary	152
9.8	Rolling position	153
9.9	Grooming attitudes	155
10.1	Three-week-old fetus within the uterus	160
10.2	Birth dates of badgers from south-west England and estimated start of weaning times at three months of age	161
10.3	Cubs aged five weeks, raised in captivity	167
10.4	Frequency of long-duration matings	172
10.5	Photomicrograph of a blastocyst recovered from the uterus in June	175
10.6	Examples of variation in the period of delayed implantation in the badger caused by February, May and autumn matings	177
11.1	Distribution of badger setts (occupied and unoccupied) in England, Scotland and Wales (absent from Orkney and Shetland) with reliability diagram inset	183
11.2	Three common badger ectoparasites. 1. Louse, *Trichodectes melis*; 2. flea, *Paraceras melis*; 3. tick, *Ixodes hexagonus*	189
11.3	Badger road casualties, 1973–6, examined at Gloucester Veterinary Centre	193
12.1	Many badgers are killed crossing roads	197
12.2	Stream culvert of the Armco galvanised steel type under a by-pass in Hertfordshire, showing raised badger path	198
12.3	Badger-proof fencing. a) Timber post-and-rail fence with wire dug in; b) wire laid flat on surface of ground, lightly turfed	199
12.4	Badger tunnel built under the M5 motorway in Somerset	199
12.5	Badger using the type of swing gate recommended by the UK Forestry Commission	206
12.6	Badger gate specifications	207
12.7	With regular feeding and patience, wild cubs may be persuaded to feed from the hand	208
13.1	Distribution of the four sub-species of the American badger, *Taxidea taxus*	210
13.2	American badger, *Taxidea taxus*	211
14.1	General range of ferret badgers, South-East Asia	218
14.2	Burmese ferret badger	218
14.3	Honey badgers	220

Tables

4.1 Percentages of setts in various habitats in selected counties of Great Britain — 68
4.2 Number of setts in various geological strata in Sussex. From Clements (1974 updated) — 70
5.1 Mammalian tenants of setts — 80
10.1 Growth rate of female badgers in the wild — 166
11.1 Territory size, group size and badger density in four areas — 184

Acknowledgements

I would like to express my great thanks and indebtedness to all those whose writings I have consulted; in particular to Martin Hancox, who kindly allowed me to draw freely from his thesis; to my son, Keith Neal and Dr Roger Avery, who made available the results of their remarkably sustained observations at a sett in Somerset over four years; to Mrs Chris Ferris who has generously allowed me to quote from her diaries some outstanding and unique observations of badger behaviour; to George Barker and Dr Keith Bradbury, who allowed me to use unpublished material on the food of the badger; to Professor R.J. Harrison, FRS for indispensable assistance when researching on the badger's reproductive biology; to Dr Paul Benham for permission to quote from his unpublished research on cattle/badger interactions; to Dr John Gallagher for most useful data on badger road casualties; to Dr Stephen Harris for supplying data on urban badgers; to Dr Hans Kruuk for his most valuable contributions to behavioural and ecological research; to Professor Roy Anderson and Dr Wendy Rees for most helpful comments on certain sections of the book and especially to Dr Chris Cheeseman for kindly making available the fruits of his most valuable researches and for commenting on particular chapters so helpfully.

I am also most grateful to Michael Clark who took so much care over the artwork with such attractive results and for his interest in the whole project. I also thank most warmly all those who contributed to the Mammal Society's National Badger Survey and in particular to Clem (E.D. Clements) who took endless trouble in analysing the data and producing the revised sett density map, and to those photographers who allowed me to use some of their best pictures, as acknowledged below.

I also have much pleasure in thanking the vast number of correspondents who have written to me over the past 35 years about their experiences with badgers. I count myself fortunate in having acquired so many friends through this mutual interest.

Finally, I must express my deep gratitude to my wife; without her patience, understanding and encouragement this book would not have been written.

Illustrations

Most of the photographs in the book – colour and black and white – were

ACKNOWLEDGEMENTS

provided by the author, Ernest Neal. Other sources, which the author and publishers gratefully acknowledge, are listed below.

Ardea Photographics: Colour Plates 18 and 19. Eric Ashby: Colour Plate 17. Gordon Burness: Colour Plates 5 and 6, Figures 2.5 and 12.5. M. Chesworth: Figures 3.12, 3.15, 8.1, 8.2 and 8.12. Michael Clark: Figures 1.4 and 12.2. E.C.D. Darwall: Figure 8.9. Professor R. Harrison: Figure 10.5. Frank Hawtin: Colour Plate 15. Pat Morris: Figure 13.2. Keith Neal: Figures 3.13 and 9.5. G.A. Pawsey: Figures 3.14, 8.10 and 12.1. Harold Platt: Figure 7.2.

The line drawings, sketches and scraper-board illustrations were specially prepared by Michael Clark.

Series editor's foreword

In recent years there has been a great upsurge of interest in wildlife and a deepening concern for nature conservation. For many there is a compelling urge to counterbalance some of the artificiality of present-day living with a more intimate involvement with the natural world. More people are coming to realise that we are all part of nature, not apart from it. There seems to be a greater desire to understand its complexities and appreciate its beauty.

This appreciation of wildlife and wild places has been greatly stimulated by the world-wide impact of natural history television programmes. These have brought into our homes the sights and sounds both of our own countryside and of far-off places that arouse our interest and delight.

In parallel with this growth of interest there has been a great expansion of knowledge and, above all, understanding of the natural world—an understanding vital to any conservation measures that can be taken to safeguard it. More and more field workers have carried out painstaking studies of many species, analysing their intricate behaviour, relationships and the part they play in the general ecology of their habitats. To the time-honoured techniques of field observations and experimentation has been added the sophistication of radio-telemetry whereby individual animals can be followed, even in the dark and over long periods, and their activities recorded. Infra-red cameras and light-intensifying binoculars now add a new dimension to the study of nocturnal animals. Through such devices great advances have been made.

This series of volumes aims to bring this information together in an exciting and readable form so that all who are interested in wildlife may benefit from such a synthesis. Many of the titles in the series concern groups of related species such as otters, squirrels and rabbits so that readers from many parts of the world may learn about their own more familiar animals in a much wider context. Inevitably more emphasis will be given to particular species within a group as some have been more extensively studied than others. Authors too have their own special interests and experience and a text gains much in authority and vividness when there has been personal involvement.

Many natural history books have been published in recent years which have delighted the eye and fired the imagination. This is wholly good. But it is the intention of this series to take this a step further by exploring the

SERIES EDITOR'S FOREWORD

subject in greater depth and by making available the results of recent research. In this way it is hoped to satisfy to some extent at least the curiosity and desire to know more which is such an encouraging characteristic of the keen naturalist of today.

Ernest Neal,
Taunton

Foreword

by Sir David Attenborough FRS

The badger is one of the larger wild animals in Britain. With its handsome striped head, it is certainly one of the most spectacular. And over much of the country it is quite common. Yet because it is nocturnal, few people ever see one. So until quite recently it seemed not to be a vivid personality in the imagination of the public, as the fox, the rabbit and the squirrel have always been.

But that changed in 1948. In that year, a Somerset schoolmaster named Ernest Neal published a book about badgers. It was full of fascinating new information and it carried the clear implication that the author had spent more time watching badgers in the wild than any other man alive. But more was to come. Soon afterwards he appeared on television with Humphrey Hewer and showed film they had taken of a badger family. The badgers were totally oblivious of the camera, the young playing boisterously with one another, the boar and the sow clearing out the bedding from their dens and replacing it with new. No one had ever seen badgers in this way before. During the programme, Ernest and Humphrey explained how they had obtained these remarkable shots. They had managed, slowly and patiently, so to accustom the animals to artificial lights placed around their burrows that eventually they came to ignore them altogether. The news that it was possible to do such a thing spread swiftly. Soon, naturalists in woods all over the country were holding badger-watching parties. It was almost as though a new species of mammal had suddenly been added to the countryside. And at last the badger acquired the hold on the public's affections that it had always deserved.

That was nearly 40 years ago. Since then, of course, Ernest Neal has continued his work with badgers — though anyone who reads what he writes will quickly realise that, for him, it is not so much work as a deep and addictive pleasure. This book is a summation of all he has learned about the species, both from the researches of others and from his own observations. It is likely to remain the authoritative account of badgers from a very long time to come.

A few people only, for varying reasons, have become so closely identified with a British animal that the name of one instantly recalls the name of the other: John Peel with the fox; Thomas Muffet, the seventeenth-century doctor naturalist whose little daughter so injudiciously sat on a tuffet, with spiders; and Sir Peter Scott with wildfowl. The name Ernest Neal is linked permanently with badgers. The pages that follow show just why.

Preface

This volume is based on my book, *Badgers*, published by Blandford Press in 1977. Long since out of print, I have now revised it considerably, brought it as up to date as possible and included much new data.

During the past decade or so much has been learnt particularly about the behaviour, ecology and population dynamics of the Eurasian badger. Some of this research was initiated through the need to assess scientifically the possible implication of the badger in transmitting bovine tuberculosis to cattle and to provide the necessary understanding for controlling the disease. It is ironical that much of this excellent work might never have been attempted but for this complex and distressing problem.

Badgers have now acquired a rather special place in the affections and interests of a much wider spectrum of people than ever before. Through television their charming and intriguing behaviour has been enjoyed by millions thanks to the patience and skill of such outstanding cameramen as Eric Ashby. Many viewers will also have nostalgic memories of the series of programmes, 'Badger Watch', when, thanks to the sophisticated techniques of the BBC, badger behaviour was shown 'live' using remote-controlled, infra-red cameras.

But many people have not been content with armchair viewing and have gone out of their way to watch badgers for themselves. In this way they have discovered the thrill of watching wild mammals at night and experienced the deep satisfaction of an intimate contact with the natural world.

I have now studied badgers in Britain for nearly 50 years and it has given me much pleasure to see how many other scientists and naturalists have caught the badger-watching bug and contributed so greatly to the understanding of these animals. It is the aim of this book to attempt to synthesise the extensive knowledge which is now available.

The world of badgers is in some ways analogous with the human world. It is made up of large numbers of individuals grouped together in discrete social groups, each badger having its own distinctive personality and temperament. Like us their behaviour is greatly influenced by their needs for homes and living space and being social creatures like we are, they too have their problems of learning how to live together ... and with us.

Ernest Neal
Milverton, Somerset

FOR
Betty,
Keith and Ruth,
David and Rachel, Andrew and Ghyslaine

1 Badgers: studying them in the field

Badgers are members of the weasel family, the *Mustelidae*, a group of mammals, typically with rather long bodies carried on short legs. They all possess musk glands which produce secretions, usually strong smelling and often used defensively or for communication.

This very successful family of small to medium-sized carnivores contains a number of familiar species which illustrate very well the evolutionary principle of adaptive radiation. Thus the ancestral forms were less specialised terrestrial forest dwellers, but during the course of evolution they diversified greatly into highly specialised species now exploiting a wide range of habitats.

Living in Western Europe today, the more typical mustelids are terrestrial, such as the weasel (*Mustela nivalis*), stoat (*Mustela erminea*) and polecat (*Mustela putorius*) which show a progression in size and so fill different niches because the size-ranges of their prey are different. A less familiar terrestrial species, the much larger wolverine (*Gulo gulo*) has become adapted to the more severe conditions of Northern Europe.

By contrast, the pine marten (*Martes martes*) has become largely arboreal. It is a marvellous climber and acrobat, being beautifully adapted to life in the trees of the more northern forests. In Britain, however, it appears to be quite as much at home in rocky and more open habitats. The very similar beech marten (*Martes foina*) has very comparable adaptations, but it does not compete much with the pine marten as it favours deciduous woodland and more cultivated land. Although its range overlaps the pine marten's, it tends to have a more southerly distribution; it does not occur in Britain.

Another highly specialised mustelid is the otter (*Lutra lutra*) which has become largely aquatic. Although quite at home on land it is only when seen in water that one appreciates how perfectly it is adapted for its mode of life.

Finally, there is the Eurasian badger which has many adaptations for a fossorial life, spending long periods below ground in an extensive burrow system it has excavated. Feeding largely on invertebrates and being much more omnivorous than other mustelids, it competes very little with them for food.

So through this considerable diversity, a number of closely related species are able to live together in the same region by exploiting different niches.

BADGERS: STUDYING THEM IN THE FIELD

Figure 1.1 Badgers of the world

Figure 1.2 World distribution of six genera of badgers. For Melogale, *see Figure 14.1*

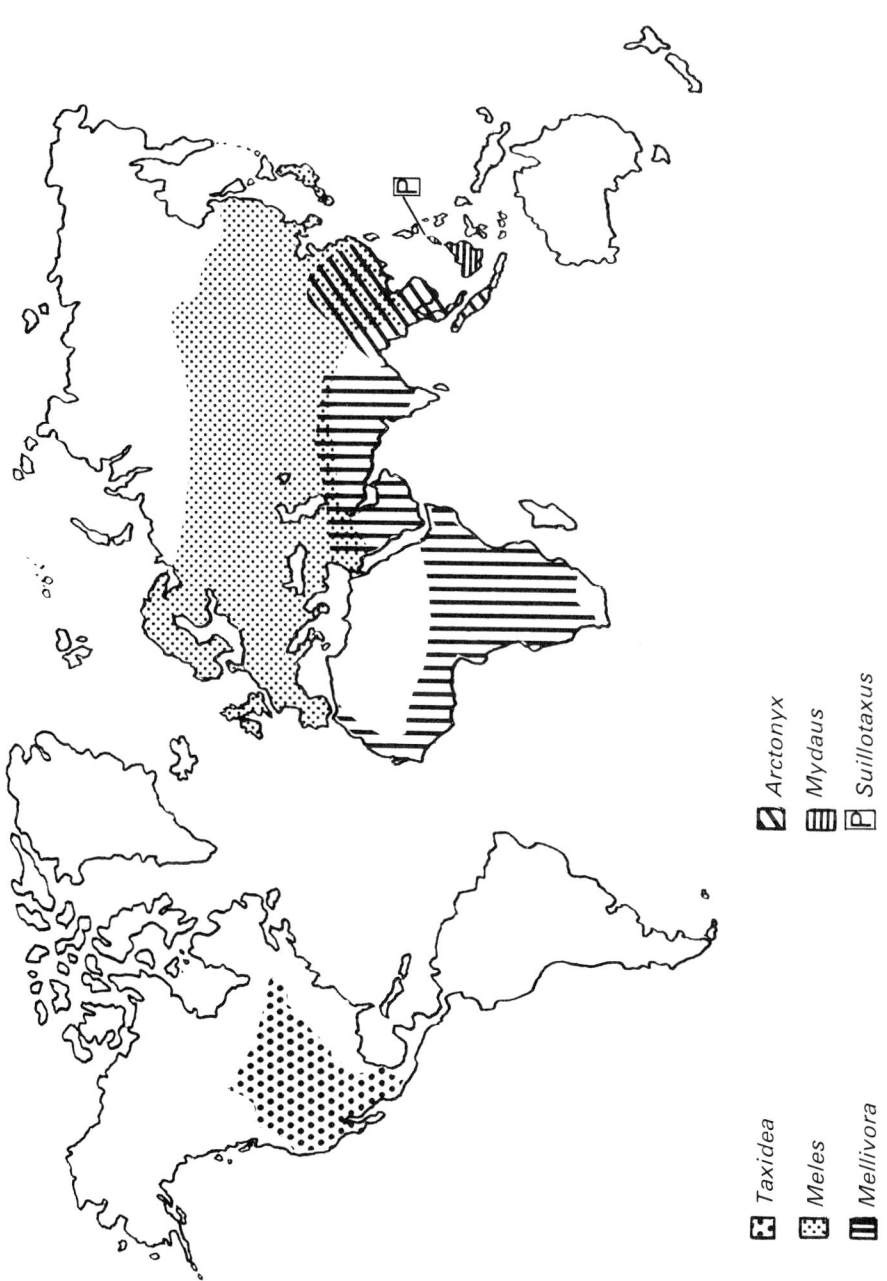

The family is divided into five sub-families: the *Mustelinae* which includes the stoats, weasels, polecats, martens and wolverines: the *Mellivorinae* with a single species, the honey badger or ratel; the *Melinae* which contains the true badgers: the *Mephitinae*, the skunks; and the *Lutrinae*, the otters.

It is the *Melinae* which mainly concerns us here, although mention will be made of the honey badger which is not a true badger, but nevertheless shows many behavioural similarities.

The true badgers are classified into six genera: *Meles*, the Eurasian badgers; *Arctonyx*, the hog badgers of Asia; *Mydaus*, the Indonesian stink badgers; *Suillotaxus*, the stink badgers of Palawan and the Calamian islands; *Taxidea*, the American badgers; and *Melogale*, the ferret badgers of Asia.

METHODS OF STUDYING EURASIAN BADGERS

The first Eurasian badgers I ever saw in the wild was in 1936. I was returning from a mothing expedition in a wood in the Cotswolds and was quietly wandering down one of the rides when I heard a great commotion going on in the wood to my right. Obviously several sizeable animals were rushing about among the dry beech leaves, occasionally making high-pitched yelps similar to those made by puppies when playing together boisterously. I tried to get nearer, but in spite of the noise they were making, they heard me and all I managed to obtain was a fleeting glance of several badgers stampeding for home. Approaching the spot, all I could see were the gaping holes of a large sett. I vowed I would attempt to see them properly the next night! That occasion turned out to be a red-letter day in my life as it started me off on a study of these intriguing animals over nearly 50 years. The following account is based on the field notes I wrote up afterwards.

> I set out in high hopes about an hour before dusk. Approaching up wind as quietly as possible, I crept under a box bush and lay down on the ground some 8 m from the main holes. It was not a good vantage point as there was a ridge which obscured my view of the larger entrances, but I dared not go closer as the light was still good. Everything was quiet except for a few birds returning to their roosting trees. A persistent cuckoo stuttered as it flew over the wood 'cuck-cuckoo', 'cuck-cuckoo', and a pair of carrion crows noisily advertised their return to a fir tree.
>
> Dusk was deepening when my attention was riveted by a loud scratching noise going on just over the ridge. It was the unmistakable reaction of an animal to its parasites and I knew that at any moment I should get my first proper sight of a badger. Within a minute or so the scratching ceased and a grey form came into view. The badger was clearly visible as it set off at a steady trot towards its feeding grounds.
>
> I hoped this was not the end, so I kept quite still and before long had my reward. A little striped face looked over the ridge, and then another; then both disappeared again to the accompaniment of a loud yelp as one cub playfully bit the other. As it was now nearly dark, I cautiously came out from under my bush, and carefully raised myself on to a bough about a metre up a lime tree. This gave me a good view of the sett, and with my back to the trunk I was not noticed. I shall never

forget that scene! Instead of seeing two cubs as I expected, there were five, and soon they were all romping round the entrance of the sett and tumbling over each other as they bit and growled. They were quite small, no bigger than large cats, and they were as playful as kittens. An adult was there too, probably the mother; she sat on one side as the cubs played together. I presumed that it had been the boar that I had previously seen going off for food. A new noise broke out when one cub found an old treacle tin, in which a stone had become lodged and for a long time they pawed and leapt on it, fighting hard for possession and giving out a series of excited yelps.

Then the noise subsided and the work of the evening commenced. They left the immediate vicinity of their home and rooted among the leaves and stones for food. They seemed to be everywhere, and I wondered what they would do if they came my way. It was not long before I knew, as one inquisitive cub came snuffling in my direction, pushing its sensitive snout into every patch of leaves. I held my breath as it passed immediately below my feet without an upward glance; its head no more than 60 cm from my shoes. But finding it was on its own, it ran off towards the others and gradually the noises grew fainter as all the badgers moved off through the wood. It was now possible to relax, so I came off my perch and left them to their feeding.

Since that time I have watched badgers many hundreds of times and have learnt much more about the best techniques for watching them, but no occasion has given me more of a thrill than that first introduction to the world of badgers.

Badgers are largely nocturnal, although in the summer months when nights are short they may emerge in the late evening and return home well after dawn. During the day, they normally remain below ground in a complex of tunnels and chambers of their own making known as a sett. Occasionally, they will make use of artificial drains put down for foxes or hide up in natural fissures in rocky areas, and exceptionally they will lie up in dense vegetation above ground: but these are usually temporary refuges.

To aid observations, binoculars with wide apertures, such as 7×50 are ideal, as in poor light they concentrate what light is available and you can see details which are quite invisible to the naked eye. In good moonlight, it is surprising how much can be seen.

After dark a torch may be used successfully. For many years I have used quite a strong one fitted with a red filter. Badgers take very little notice of the colour red and you can see most details with such a light from 10–20 m. I remember once using a lantern torch which I placed on the ground so that its powerful red beam was directed on to the sett. This was to enable me to judge the right moment to take a photograph. To my surprise one badger came towards the light, which was about 1 m from where I was sitting, and sniffed it inquisitively before moving off. It showed no alarm even though the light was shining into its eyes.

You can also use a normal torch if not too strong, but when first switched on it is best not to point it directly at the animal. Badgers take far less notice of torch light coming from above than from ground level. Perhaps they associate light from above with moonlight and so are less concerned. If a torch is used, it should be one with a silent switch.

Badger watching is not always easy. On some occasions everything goes marvellously, the badgers have no clue that you are there and you witness a wonderful uninhibited display of activity. But at other times the badgers are slightly suspicious and they go off quietly without doing much of interest. Every badger watcher has a few blank nights, and however much care is taken the badgers' suspicions are bound to be raised on some occasions, but by taking certain precautions disappointments can be minimised.

Badger Watching by a Sett

The choice of sett to watch is important; watching some setts can be extremely difficult. If you have a choice, choose one which is obviously well used, but has not too many entrances. I know of several setts in the southwest of England where there are more than 50 entrances spread out over a large area. Such setts are very difficult to watch as it is not always easy to know which entrances will be used, and when the wind is right for one hole it is wrong for another, so you are likely to be discovered. Also, it does not always follow that there are more badgers living in a large sett than a small one. In the Cotswolds, I once watched a sett which only had one entrance and out of it came twelve badgers in succession. It was like watching a conjurer producing rabbits from a hat!

Choose a sett where there is not too much undergrowth around the entrances so that you may obtain an uninterrupted view. The presence of bushes or trees within 5–10 m from the main entrances are a great help in giving you some cover.

Setts in some districts are much easier to watch than in others. This may be because the badgers are less disturbed by people or animals, but it may also be due to the terrain. In my experience, a sett is easier to watch when it is in a drier situation such as in chalk or limestone districts. In damp, ferny places where there are plenty of elders and nettles, your scent carries too well and watching is more difficult.

When going on a badger-watching expedition, some advanced planning is helpful. It is not wise to go on the spur of the moment to a sett you do not know well, but is far better to make a thorough reconnaissance first. I like to do this in the morning, as this gives time for the scent I may leave around the sett to disperse before the badgers emerge in the evening. But to make quite sure, I try to avoid standing in front of any of the used entrances or on any of the main badger paths.

First find out if the sett is occupied, because badgers often have more than one sett on their territory and a sett may be left unoccupied for weeks or months at a time.

Fresh earth thrown out of an entrance is a good sign of occupation, but I have known badgers spend a night excavating a sett and then go back to sleep in an alternative one elsewhere; so you cannot be absolutely certain. Fresh pad marks just inside the sett entrance are useful clues to occupation and if you can find fresh dung within 10 m or so from the sett, this is usually conclusive.

When a sett has many holes you need to find out which ones are likely to be used for emergence. You can usually eliminate entrances half filled with old leaves or which have cobwebs across. A well-used hole is often a large one with rather polished edges. Also the presence of flies going in and out is an excellent sign that badgers are not far below ground. The flies smell the badgers below and instinctively fly in that direction, but when they get into the semi-darkness of the tunnel another instinctive action causes them to fly towards the light, so they come out again. This shuttling to and fro can go on for a long time. It must be very frustrating for the flies, as they never quite make it!

Having decided which holes to watch, next choose your vantage point. This should finally be decided according to the direction of the wind when you arrive in the evening, so alternatives should be considered in case the wind should change. I much prefer to watch from a tree 5–7 m from the main entrances; then if the wind changes and you feel it on your back, your scent will probably blow over the heads of the badgers and you will not be detected. Trees have the additional advantage that badgers in their wanderings are unlikely to discover you, a hazard you have to reckon with if you are on the ground. The disadvantages of a tree are having to climb it in the first place, and then getting down again in the dark some time later when your limbs may be cold, stiff and cramped. It all depends on the tree! A short, light ladder may be used very successfully and it may even be possible to pull it up after you once you are in a good position. A ladder is well worth considering for regular watching. I have not yet discovered a tree which could be described as comfortable after the first quarter of an hour, but some are certainly better than others. Standing on a thin bough can give your instep agonies after a time, but sitting on a foam cushion placed on a thick one is far better, unless you fall asleep!

If a suitable climbing tree is not available you can sit or stand with your back to a large tree trunk or dense shrub. In this way you are not silhouetted against the sky. But choose a place which is not too near one of the main badger paths to minimise the risk of discovery.

I have never used a hide because, in most places, it is quite unnecessary, but other watchers have found one useful. If a hide is used it should be constructed from natural objects in the vicinity and built up gradually. I have often used branches of elder to help camouflage my camera and to break up my own outline in case the badgers should emerge in good light. Elders are particularly recommended for this purpose as their scent is strong and familiar to the badgers and may disguise to some extent any human scent.

Having made a reconnaissance, the next thing to decide is when to watch. The easiest option is to go in the evening, but watching before dawn can be extremely rewarding ... though you have to be either a poor sleeper or extremely good at getting up. For early morning watching, you should get into position a couple of hours before dawn. When you arrive the badgers will probably be away from the sett area, so watching from a tree is very desirable as this reduces the likehood of discovery when they return. Badgers usually come back about an hour before dawn, but in summer it may be in good daylight (see p. 106).

For evening watching, you need to get there by sunset at the latest, unless you know very well the habits of the particular badgers you are intending to watch. At some setts they emerge much earlier than expected and if you leave your approach too late you are liable to disturb them on arrival.

It is always worthwhile approaching with the wind in your face, and as quietly as possible, and to pause some 50 m from the sett. If the sett can be seen from this distance, so much the better. If not, you can sometimes hear badgers, especially cubs, if they are out early. If already above ground, either wait for them to go down again, or make a slight clicking noise with your tongue. This is sometimes just enough to make them go underground without alarming them unduly. Then you can quickly reach your vantage point before they re-emerge.

When you arrive at your watching position check the wind direction carefully. Cloud movements can seldom be relied upon as they may not be related to the local wind currents at ground level. Watch the smoke from a blown-out match to make quite sure. If you find the wind is to any extent behind you when facing the sett, it is best to change to an alternative position as quickly as possible.

Some setts are particularly difficult to watch because of local wind conditions. If the entrances are at the base of a cliff or quarry face, the wind is nearly always wrong if your watching point is in front. This is because of the cyclic movements of air in such positions with an updraught from the base of the cliff. Local wind conditions also often change at dusk, especially in summer and in places where the sett is on the side of a valley. This can be exasperating, as when you arrive the wind may be slight and blowing in your face, but at dusk the warmer air in the valley rises and so there may be a complete reversal of direction for half an hour or more just when you may be expecting emergence.

It is important to wear suitable clothing. Badger watching even in summer can be a cold occupation when you have to be still for long periods; so clothes should be warm. They should also be drab in colour and not rustle when you move, or have a pronounced odour.

The most critical time for the watcher is when the badgers are about to emerge and just after. They often come near to an entrance and wait for a short time while scenting and listening. You may not be in a position to look right down the tunnel and see them, so it is always best to be as quiet as possible. When badgers are above ground it is also necessary to keep still as movement is quickly detected. If a mosquito settles on your face when a badger has just emerged, you should avoid making a quick movement to remove it, but it can be done in slow motion! Midges and mosquitoes can be a real menace in some areas and it is a useful precaution to cover up as much as possible of your anatomy except for your face. Two pairs of socks are better protection for your ankles than one. Some people anoint themselves with anti-midge cream; if the wind is right, this does no harm, but otherwise I prefer not to use it.

From these remarks it might be judged that watching badgers successfully is a most unlikely event. This is not so. On some occasions you seem to be able to get away with anything. The wind may be variable, but the badgers are not alarmed; you make noises and they take no notice; you may even wave your arms about and a cub may be so inquisitive that it comes right up to you to investigate this strange happening! But usually this does not happen, and if you wish to avoid disappointment it is far better to take as many precautions as possible, especially at setts which you have not watched previously. If you watch a sett regularly, the badgers may get used to your scent and become remarkably tolerant of your presence.

When leaving for home, it is advisable to wait until all the badgers have left and then go quietly, without passing too near the entrances or treading on main badger paths. This makes future watching more likely to be successful as the badgers will not have associated disturbance with any scent which may have lingered around the sett.

Following Badgers After They Have Left the Sett

Much can be discovered about badgers by watching regularly at a sett, but it is always rather frustrating to see them go off into the darkness and be unable to follow them. What they do for the rest of the night is more difficult to discover. You can find out a lot by tracking during the daytime, especially in snow; this may tell you where they went, and sometimes how and on what they fed, but most of their behaviour is never understood by this means and for most of the year any form of tracking is of limited use.

It is possible to follow them after they have left the sett, but it is difficult. You have one advantage; badgers do not usually wander about aimlessly but follow their main tracks for some distance before starting to feed, so you can follow these trails some way behind the badger until it starts hunting for food. Then it may become more difficult.

You need to walk very quietly along the path used by the badgers on leaving the sett. Soft-soled shoes are ideal for this. A weak torch will help you to avoid stepping on twigs. Stopping to listen at frequent intervals is a necessary precaution as you don't want to come upon the badgers unawares. They are often noisy in their movements, especially if there are stones or dry leaves about and if you cannot move quietly yourself, it is best to imitate the sort of noises the badgers are making. When badgers are moving quite quickly they do so in a series of short trots, stopping at frequent intervals to listen. It is not difficult to imitate this movement. If they become suspicious, it is a good idea to stop and make the scratching noise a badger makes so often after emergence from the sett. This can be done by rubbing your fingers rapidly against your neck or clothes. This often has a useful calming effect on badgers as they usually only scratch when in an unsuspicious state.

Once badgers are feeding they are not so alert, and it is possible with a torch to stand still and watch them catching earthworms, turning over cow pats for beetles and searching elsewhere for other delicacies.

Following badgers is much easier when the ground is soft after rain and you can be more silent, and when the wind is right for following the badger you can approach closer. It is also a help when there is just enough moonlight for you to see where you are treading without using a torch, but is not too bright to make you conspicuous. You need some luck, but you improve with practice, and occasionally this technique can be very successful.

It is also possible, with experience, to go to particular places where badgers are likely to be seen and wait for them to arrive. Favourite feeding grounds are the most rewarding, but there is a considerable degree of chance with this method unless you know your particular badgers extremely well.

But by far the best method is to habituate badgers to your presence so that they accept you as part of their world. This has been done with outstanding success by Mrs Chris Ferris who is the first person to my knowledge to exploit this technique with wild badgers. Night after night for many months she was 'just there'. At first the badgers avoided her, later they became curious but kept their distance, but finally they accepted her presence completely with remarkable trust. By setting scent on her they claimed her as one of their social group and came over to greet her when she appeared. In this way she was able to make unique close-up observations of their activities, some of which I am privileged to include in this book.

BADGERS: STUDYING THEM IN THE FIELD

Research Methods

The scientist has now got other methods of following badgers by night, the chief being telemetry. For this work, the badger has to be caught in a live trap, anaesthetised briefly and fitted with a harness containing a minute radio transmitter. This gives off a continuous series of signals which can be picked up as bleeps by a receiver carried by the research worker. In fairly flat country, these signals can be detected up to 1 km away. This technique, theoretically, enables you to find a badger at any time after it has left the sett. By pointing the aerial in various directions you can find out from which point of the compass the strongest signals come: this gives you the direction in which the badger may be found. In addition a beta light reflector may be incorporated in the harness which glows when light shines on it. This makes it easier for the observer to locate the actual position of the animal. By using separate harnesses on different badgers of known age and sex, you can follow the activities of each individually, as each transmitter can be tuned to give a different set of signals.

There are also available special binoculars which intensify light to such an extent that the animal can be watched in excellent detail, even on fairly dark nights. However, in practice they have one drawback as light-coloured objects and light from the much brighter sky or from the moon give such contrast that in some situations the badgers you are trying to see are quite difficult to observe. Better binoculars for the purpose are available which use infra-red light. Fixed above the binoculars is a powerful lamp fitted with a filter which only allows the passage of infra-red rays. They emerge as a parallel beam of 'invisible light'. The lamp is powered by a battery pack slung on your back, and when the binoculars are pointed at an object in the dark, the reflected infra-red rays are converted within the binoculars into visible light, so you can literally see in the dark. Thus, telemetry may be used to locate the badger and then the binoculars to observe its behaviour.

It is an incredible experience to follow the activities of badgers in this manner. I greatly enjoyed the chance to do this with Christopher Cheeseman in the course of his researches in Gloucestershire.

One such occasion was in September. The badgers were given plenty of time to leave their setts and reach their foraging areas before we started out to track them. Three badgers from different social groups had previously been fitted with harnesses containing radio transmitters, each giving a different signal. We stopped at strategic places to try to locate the position of the particular animal we were looking for, having tuned in to its personal wavelength. A faint bleep indicated that our quarry was several hundred metres away and the position of the aerial when the sounds were loudest gave us the direction to follow.

It happened to be a particularly dark night on this occasion and as one stumbled about in the darkness over country littered with unsuspected hazards one envied the badger's ability to find its way unerringly using its sense of smell and hearing. Fortunately, cows lying on the grass were obstacles which could usually be avoided as they could often be heard chewing the cud or snuffling before one walked into them. We used a small torch when absolutely essential but preferred not to do so when nearing our badger. The best plan was to proceed very slowly, checking direction at intervals by sound signals from the receiver.

A direct approach is not always advisable as it is important to be down

Figure 1.2 Radio tracking of badgers. a) Badger fitted with transmitter; b) harness and transmitter. Battery life: 14 months (small size) and up to three years (larger size). Range about 1 km. All parts are non-corrosive

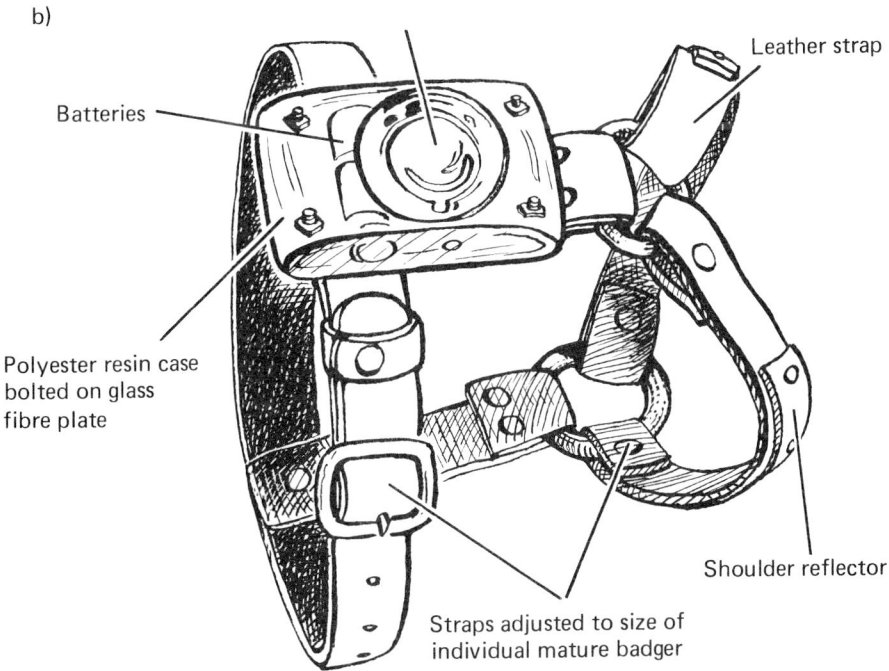

BADGERS: STUDYING THEM IN THE FIELD

Figure 1.3 Apparatus for the location and observation of badgers wearing transmitters. a) Aerials, receiver and earphones. The 'H' Adcock aerial uses a 360 compass scale, for two operators to take triangulation bearings from two or three points; b) viewer, power pack and the binocular viewer which uses an infra-red beam over a single central lens

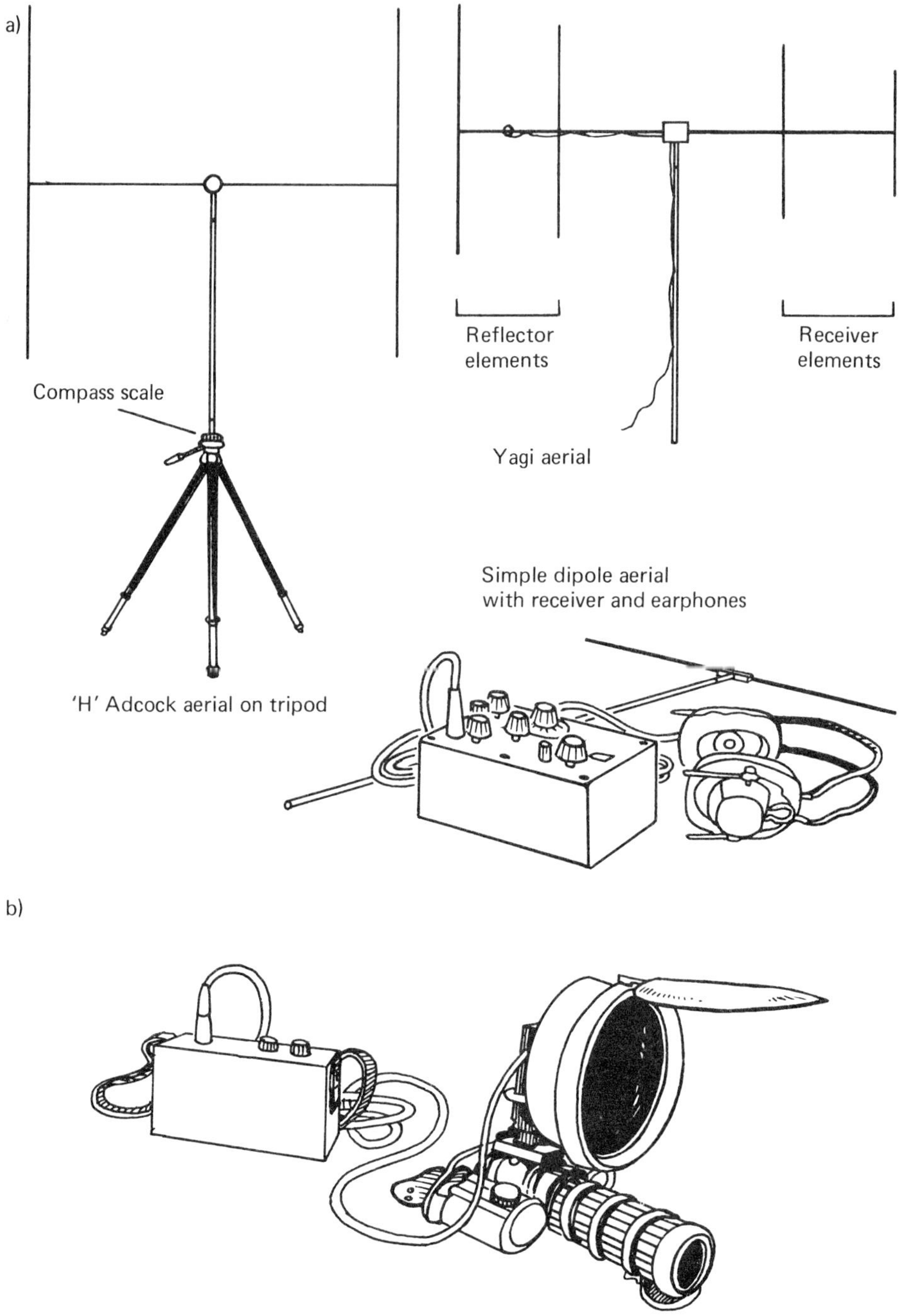

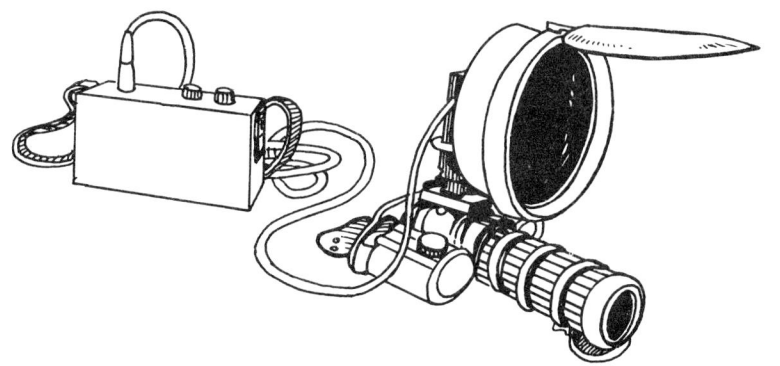

Figure 1.4 Watching badgers at night using infra-red binoculars

wind when you get fairly near. With a steady wind in your face you can literally walk right up to a badger if you are quiet, but with variable wind you have to be content with more distant viewing.

Scanning the field where the badger was thought to be foraging we soon picked up the beta light reflector on its harness and, with the infra-red binoculars, could watch its activities as if it were daylight. We had to remember that other badgers might be about which had no radios to tell us of their position, so careful scanning of the meadow with the binoculars was essential before trying to get closer. It was thrilling to be able to watch a badger in almost complete darkness with the animal quite oblivious of being watched. This technique certainly brings a new dimension to badger watching! On many occasions Cheeseman has been able to stand quite close to a badger and not only identify what was being eaten but also to count the individual items.

Unfortunately for the amateur naturalist, the apparatus concerned with telemetry, and infra-red binoculars in particular, are extremely expensive, but as research tools they are invaluable.

2. General characteristics

EXTERNAL FEATURES

The characteristic features of a badger are best considered in relation to its fossorial mode of life. It was Harting (1888) who first suggested that the word badger came from the French word *bêcheur*, which means a digger. If this is true, it is certainly aptly named, because its digging activities are quite phenomenal.

A badger is very powerfully built. It has rather a small head, a thick, short neck, a rather long and wedge-shaped body and a very short tail. The body is carried on short, but extremely strong, limbs and the feet armed with very strong claws. These characteristics all help to make the animal a most efficient digger, capable of working heavy material in confined spaces.

The badger has a rather elongated snout which it uses when rooting much in the same way as a pig. The end of the snout (rhinarium) has an almost rubbery texture, and being very flexible, the badger is able to peel it back away from the upper jaw when probing for food or digging in the ground. This lessens the risk of injury to this very sensitive area and prevents particles getting up the nostrils.

It would appear from the anatomy that it is capable of closing the nasal canal by muscular action about 5 mm from the openings of the nostrils. It is not easy to see if this happens in a live badger, but if it does, it would certainly serve a very useful purpose during digging operations. Nevertheless, in dusty conditions, a badger will repeatedly blow out air through the nostrils in short blasts to clear away particles taken in during essential breathing operations.

Vibrissae (whiskers) are present on the snout. These are long, stiff, black hairs with a supply of nerve endings at the base. They act as tactile sense organs. In the badger they are shorter and far less conspicuous than those of the otter, cat or rabbit. As with other animals which pick their way through narrow spaces and thick vegetation, the vibrissae probably help the badger to gauge the width of an opening. This applies especially to those which project sideways and arise from a region near the front end of the dark facial stripe. However, the four which project in various directions from above each eye probably warn the animal of anything near enough to damage that organ.

GENERAL CHARACTERISTICS

Figure 2.1 Head of an old boar showing rhinarium and whisker pattern

The eyes are small in comparison with the size of the head and appear to be of far less importance than the other sense organs.

The ears, like those of many animals which do a lot of digging, are also small and lie close to the side of the head. Although they are almost rigid, badgers can move the posterior ridges forwards, so closing the aperture. This happens automatically when digging in loose soil: a useful adaptation which helps to keep the ears clean.

The limbs are extremely efficient digging tools. Fore and hind limbs are of comparable size and their musculature is exceptionally well developed, especially in the former. A badger's immense strength may be gauged from the fact that one was seen by C. Cheeseman to tip upon end a huge stone weighing 25 kg in order to get at food he had placed underneath it!

The feet are broad and strong and have five digits, the hind feet being smaller. Both are protected on the underside by a very thick cornified layer. There are also special pads of dense connective tissue which form cushions over those joints where most pressure is borne when digging and running. These pads make the characteristic impressions of a badger's spoor and consist of five separate digital pads near the tip of each digit, and others over the corresponding joints between the toes and the metapodials. The latter are all fused together to form a single, broad, plantar pad. When walking, the wrist and heel, which are behind the plantar pad, are kept slightly off the ground in the digitigrade manner.

The claws, unlike a cat's, are non-contractile and those of the fore feet are much longer and stronger than the hind claws. It is surprising that with so much digging, the front claws become far less worn than the hind. Only in very old animals is the wear in the front claws noticeable, but it is seen in the hind claws even in quite young animals. A possible explanation could be

Figure 2.2 a) Fore foot; b) hind foot. Photographs of the same badger

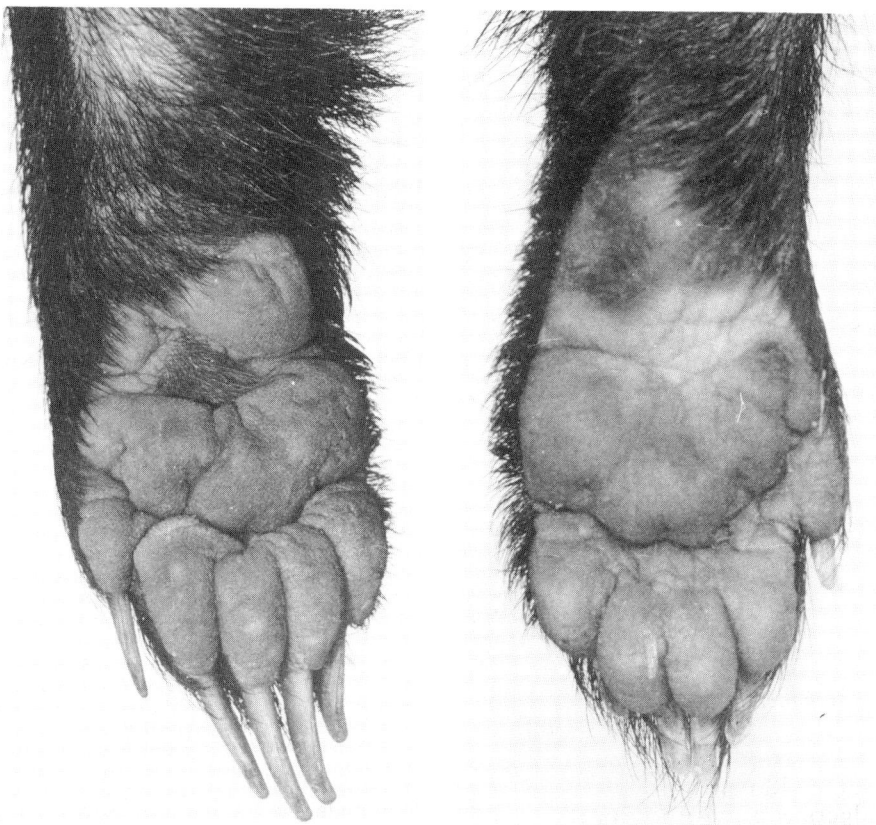

that the claws of the front feet grow faster than the hind ones to compensate for greater wear. This happens in mongooses and other carnivores which dig a lot. However, it may partly be due to the different manner in which the fore and hind feet are used during digging (p. 44).

The claws are not only used for digging, but are useful when grooming; they also act as defensive weapons. In connection with this, J. MacNally (1970) writes about a most interesting example:

> I wondered how injuries sustained by terriers when badger digging could be caused, until I saw by lucky chance, a boar badger fighting off three terriers in a cairn. He came time after time up to the rocky threshold of the cairn, driving the dogs out with lightning slashes to right and left with his long-clawed front feet. I no longer wondered how a terrier could have hair, skin and flesh torn off its head and left dangling.

LOCOMOTION

A badger's movements are related to its size and shape. It is a fairly heavy animal (p. 21), with a rather long back, so the spine is kept fairly rigid to give the body sufficient support. It is therefore unable to arch its back as

Figure 2.3 Footprint in soft mud, showing heel mark

much as many shorter animals, or pounce like a fox or a cat. As the limbs are of comparable length the back is kept horizontal when it moves, and having short legs which move quite rapidly it sometimes appears to glide along.

When walking slowly there are usually three legs supporting the body at any one time. The legs are moved symmetrically, with the left legs repeating the movements of the right legs, half a stride later. When the trail of a walking badger is followed, you can see that the tracks of the smaller hind feet are superimposed on those of the front feet, but they often do not register completely. The tracks are placed close to the median line, but not so exactly as a fox.

When a badger trots, its body is supported by two diagonal legs at one time. This is the most common form of locomotion when a badger is going fairly rapidly from one sett to another or to its feeding ground. When trotting, the head is held low down and swings gently from side to side, the hind quarters swaying in sympathy. Trotting, however, is not continuous as badgers stop at frequent intervals to listen. The trail made when trotting shows the tracks further apart than for a walk and they are usually not quite registered.

When a badger is really in a hurry, it will gallop. It does this when badly frightened or in pursuit of another after a territorial dispute. When galloping the legs are moved asymmetrically and, at times, the body may be completely unsupported. A badger galloping in front of a car kept up a speed of 25–30 km per hour over a short distance. They certainly cannot keep up a gallop for long as R.H. Mallinson mentioned in a letter to *The Field*:

> At about 22.30 on a moonlight night I suddenly became aware of a most extraordinary noise. At first I thought it was an engine with a leaky boiler on the distant railway line, but I quickly realised that it was close and drawing near. Then into view came two badgers one behind the other. They were running as hard as they could and wheezing and blowing as if in the last stages of exhaustion. They pounded past me and through a wire fence. Badger number one continued on into the wood, but badger number two suddenly collapsed. He stretched and rolled from side to side on the dewy grass, exactly like a dog, puffing and wheezing all the time. I slipped through the fence and turned a powerful torch on to him, but he continued to roll about until I was 10 feet from him, when he stopped blowing, stared at me intently, scrambled to his feet and, after a moment's hesitation, lumbered away.

The trail made by a galloping badger shows the tracks well spaced out and not registered. When interpreting the tracks of badgers, it should be borne in mind that the print of the fore feet are at least 5 mm broader than the hind, and that in the fore foot the inner toe is set further back. The marks of the claws are usually conspicuous, especially in the fore foot, and their points are further away from the digital pads than they are in the hind. The width of the fore foot of an adult male of about 12 kg is about 50 mm and 45 mm in a female of about 10 kg. However, much larger prints do occur, up to about 65 mm for a very large boar.

Climbing does take place occasionally although badgers are not the best size and weight to climb well. When playing they will clamber over logs and fallen trees, but their usual reason for climbing trees appears to be to find slugs. On wet nights slugs often browse on the lichens and algae on a tree trunk, and if the bark is rough enough, as in the case of elders, a badger will climb like a bear, clasping the trunk with both fore and hind limbs and getting a grip with its claws. The mud from the feet and the marks of a badger's claws may sometimes be seen up to a height of about 5 m, but this only happens when the trunk is sloping. A vertical trunk is never climbed successfully, although occasionally one may make an attempt. Badgers sometimes use fallen trees as bridges to get across a stream or ravine.

Occasionally, badgers have been known to fall into swimming baths which have been emptied for the winter. M.R. Dunwell described such an incident which occurred in February 1959. The small swimming pool was in an orchard, and the badger was found one morning curled up in a pile of leaves in one corner. Some rough planks which had been nailed together were placed up the side as a ramp, but the badger could not be persuaded to use them, but going to one corner it tried to climb the vertical wall. Slipping back, it tried the reverse procedure and stood on its front legs with its underside facing the corner and tried to climb up backwards. It even stood on one front leg so as to have three legs working at the wall. It could not quite manage it, and when left alone it soon retired to its nest of leaves. It later escaped unseen, presumably with the help of the planks.

In February 1963 there was a rather similar instance when two children, P. Brinsmead-Williams and S. Dickins discovered a badger in an empty swimming pool at Pill, Avon. Although the pool had been drained it had up

GENERAL CHARACTERISTICS

Figure 2.4 Badger climbing an elder tree after slugs

to 150 mm of water and ice over three-quarters of its floor. The badger, having no doubt accidentally fallen into it the previous night, was curled up in the driest corner. A long plank was arranged as a ramp and they tried to encourage the animal to freedom. Its first approach to the ramp was in reverse and it backed up about a metre before losing its footing. Eventually, induced to try a more orthodox direction, it went trundling up the plank and out of the pool and away. A short film of this saga was shown on television by the BBC after the news bulletin!

It is of interest that in both these instances the badger tried to go up backwards. This appears to be the usual method when climbing is difficult.

Badgers can also swim well, although usually they prefer to cross a stretch of water by other means. I was told that badgers which occupied a sett near the River Yeo in Devon would regularly come out at dusk, make their way to the river, 'belly flop' into it and swim across to the other side where there were good foraging grounds.

Mark Fisher also related to me how his tame badger, much to his surprise, swam straight across a tarn in the Lake District, Cumbria, at a point where it was about 40 m across. It held its head high, snorting quite a lot, and swimming with the usual dog-paddle technique. Others who have kept badgers as pets have commented upon their liking for water and their ability to swim well.

Sylvia Shepherd (1964) relates how her tame badger 'Brocky' took a bath most nights in hot weather. It would lie on its back across the beck damming it up until the water flowed over its tummy. It would use the same place each evening, but would never use the lake for that purpose.

GENERAL CHARACTERISTICS

Major Seale, the Coastal Warden for the National Trust at Newquay, Cornwall, had a tame badger which was a marvellous swimmer. It loved swimming in the sea and became quite a local tourist attraction in consequence.

WEIGHTS AND MEASURES

A sample of 31 badgers from Somerset which I measured and knew to be over a year old showed that the average head-plus-body length of males was 753 mm, range 686–803 mm and the average tail length was 150 mm, range 127–178 mm. In females, the corresponding averages were 724 mm, range 673–787 mm, and 150 mm, range 114–190 mm. The average total lengths came to 903 mm for males and 874 mm for females.

Larger animals from southern England measured by S. Fargher reached 880 mm plus 160 mm tail, total 1,040 mm for a male, and 880 mm plus 130 mm tail, total 1,010 mm for a female. However, some smaller animals had much longer tails than the above including a male and female both with tails of 200 mm.

Badgers are heavy for their size, but weights vary considerably according to the available food in the district and the time of year. In a sample of 117 animals from southern England the average weight of males known to have been over a year old was 11.6 kg, range 9.1–16.7 kg, and for females of a similar age 10.1 kg, range 6.5–13.9 kg. The wide range of weights is mainly due to the accumulation of fat in late summer and autumn giving maximum weights during the October–January period. Minimum weights occur in the March–May period when surplus fat has been used up.

The greatest weight changes occur in adult sows, probably because of the long lactation in the spring; 22 sows weighed during the September–February period averaged 12.2 kg, while 25 weighed between March and May averaged 8.8 kg. One sow of only 6.7 kg had been suckling cubs at the time of her death.

Again, the loss of fat during the spring period varies greatly according to the severity of the winter. In mild, damp winters in the south-west of England, badgers are active on most nights, except for a brief spell in December when they feed very little (p. 104). Under these conditions they can replace the fat used up by the food they eat. But when winters are cold and springs are dry, food is difficult to procure and weights can drop to very low levels.

Weights also vary a good deal according to the region. For example, K. Baker weighed a large sample of badgers killed by farmers during the bovine tuberculosis scare in West Penwith, Cornwall. Although Cornwall is a mild county, they averaged less than Somerset badgers by about 2 kg. Badgers were numerous in the Penwith area and it is possible that food supply was inadequate for such a high density and weights were low in consequence.

Weights also appear low in parts of Scotland. In the rather mountainous region of Inverness-shire, W. Marshall told me that in a sample of eight animals over a year old the average weight was 7.0 kg, the heaviest being 8.6 kg. It is true that these animals were all weighed in March–April, so he estimated that they would have been about 2 kg heavier in the autumn, but a low figure nevertheless.

However, not all Scottish badgers are light. One in Argyll measured

988 mm and weighed over 14 kg. It was shot as it escaped from a cart of hay! The badger's lair was in the rick which had been lifted on to the cart! Mortimer Batten (1923), a man who had much experience of Scottish badgers, always maintained that Highland badgers were heavier than Lowland ones. More data will be needed to clear up the situation.

Low weights are also characteristic of Scandinavian badgers, although Perarvid Skoog tells me they tend to put on more weight in the autumn than British ones. I was able to observe some of the Swedish badgers on one of the inshore islands of the Baltic and was very struck by their small size compared with those I was used to in Britain. Food scarcity over most of the year would again seem to be the main reason for small size, although hereditary factors cannot be ruled out.

It is interesting to note the variation in weights of Russian badgers. Although their head-plus-body lengths are comparable to those in Britain, they put on so much fat in summer and autumn that they become as round as a barrel and reach phenomenal weights. In Moscow province (Ognev 1935) the average autumn weight was 20 kg, but one killed on 23 September weighed 34 kg. That must have been quite a badger!

However, exceptionally large badgers are not confined to Russia, as in Britain a number of very heavy animals have also been recorded. These include one of 23.6 kg which was trapped at Uckfield in Sussex in 1948, and another of 25 kg, killed near Folkestone in February 1949. Brian Vesey Fitzgerald also wrote to me about three badgers taken from a sett in Durham; one weighed 27.7 kg and the other two, one of them a sow, approached that weight. In addition, A. Bramble records that a badger, killed near Rotherham in Yorkshire in December 1952, weighed 27.3 kg; it was weighed by the Scouts under the supervision of the local vicar! To what extent these exceptional weights are due to heredity or environment cannot be assessed, but food supply clearly plays a part. This is evident from the large weights attained by badgers kept in captivity when given plenty of food and usually insufficient exercise. I know of one which, at only nine months, weighed 16 kg!

THE SKIN AND PELAGE

A badger's skin resembles that of the ratel of Africa and Asia in being remarkably tough and extremely loose. It has aptly been described as 'rubbery' in texture. These properties make it extremely difficult for an attacker to grip any part of a badger's body without being badly bitten in return. Today, in Europe, there are few animals apart from dogs and the exceptional fox protecting cubs that would attack a full-grown badger. Only dogs which are large and strong can inflict much damage as the teeth of smaller ones cannot penetrate the skin at all. In fact, the skin is so loose that if a terrier does get a hold, its teeth merely grip a fold of the badger's skin.

Badgers when playing together often bite hard enough to cause a yelp, but they do not damage the skin. However, when adults fight for dominance or in defence of territory the strength of their bite is enough to make nasty wounds, especially in the region of the rump.

The pelage consists of long guard hairs which are 90–100 mm long in the winter coat, and a thick felt-like underfur. The guard hairs are characteristically pigmented with melanin which shows as a band of darker colour, about 20 mm long, rather nearer to the tip than the base. The overall effect is to give the badger a greyish look when seen from a distance. The underfur is light in colour and contains no melanin; so when the guard hairs are disturbed by wind, or when the animal is very wet, light patches show through and disturb the general greyness of the dorsal and lateral regions.

On the limbs and the ventral side of the neck, chest and abdomen, the hairs are shorter and dark all over except in some regions where there is a very short lighter portion at the base. The amount of hair on the abdomen is always far less than elsewhere, the guard hairs being short and underfur absent. In some lactating sows, the abdomen is almost denuded of hair.

The hairs of the head and neck are short and white, except for those which form the characteristic dark stripe on each side. This starts in the area behind the ear, passes through the eye region and turns down slightly towards the jaw about 20 mm from the black tip of the snout. The ears are also tipped with white hairs although specimens have been reported without these, a possible result of fighting. A strange variation of facial pattern was described to me by Michael Picken. In addition to the two facial stripes, it had a thin dark band which appeared like a bridge between them in the region above the eyes giving a somewhat triangular pattern of dark. This badger was seen on several occasions at a sett in the Mendips.

The hairs of the tail are long and lacking in melanin. This is worth remembering, because on one occasion, having found some long white hairs incorporated in lumps of clay at the sett entrance I thought I had discovered where an albino was living. It was disappointing subsequently to watch perfectly normal badgers emerge . . . but they had white tails!

When the hairs develop in a badger's skin the hair follicles do not occur singly, but the primary one gives rise to secondary follicles surrounding it. The main follicle forms a guard hair and the subsidiary ones the finer hairs of the underfur. The latter are arranged rather like the pappus of hairs on a thistle fruit.

Badgers moult once a year. Novikov (1956) describes the process as follows:

> The moult begins in the spring, the underfur falling out first, then the guard hairs. The process begins on the withers and shoulders spreading to back and flanks. The guard hairs are also shed gradually from the beginning of summer in the same sequence as the underfur. New guard hairs grow in the late summer followed by new underfur. The process continues in the reverse order from the spring moult, i.e. from the posterior part of the back to the head. The growth of the pelage ends in the autumn.

In the late summer and autumn when fat is laid down much of it is stored in the skin. It is deposited in thick layers between muscle, so in section it resembles 'streaky' bacon. It has long been thought that badger fat has special therapeutic properties, especially for the cure of rheumatism, strains and sprains. It certainly has excellent penetrating properties if rubbed into the skin. I was sufficiently interested in these reports to have the fat analysed, but this did not throw up any special substances which might be curative. However, I will relate the following story for the sake of the record. An old lady living in a village in Somerset had a badly swollen hip joint due to dislocation or disease. A bone setter told her that the only possible cure was badger grease as it was an injury of long standing. Mrs

Davies, who heard of the lady's plight, tried everywhere to find some badger grease without success, but eventually telephoned me and told her story. I had no fresh grease, but I did have some which was at least three years old which was very rancid and almost orange in colour. I hardly liked to offer it, but Mrs Davies thought it was worth trying. She rang me up a fortnight later to say that the day after the first application the skin went very red and started to draw, later the swelling went down, and by the time she telephoned me, the lady was completely cured and was overjoyed. I have often wondered since whether it was faith healing, or perhaps the action of substances formed during the bacterial action in making the fats rancid. It would be interesting to know!

The Significance of the Colour and Pattern

The colour and pattern of an animal is always of some significance. In some species cryptic colours play a major role in survival. Others have colours and patterns which warn potential predators not to attack as they may be poisonous, distasteful or have some potent means of defence. Again, colours and patterns may be used as social signals or as recognition marks between members of a species, the white patches on the rump of deer or the tails of rabbits are examples. In some animals, the colours and patterns show a compromise between two essential requirements, protection and recognition in some, or protection and warning patterns in others. In the light of this diversity of function, how can the colour and pattern of the badger be explained?

Many nocturnal animals are cryptically coloured, the majority being brown or grey. This certainly applies to the body of a badger, but certainly not to its head. When you can see anything at all in a wood at night, you can see a badger's head very plainly. There are circumstances, especially on moonlit nights, when the peculiar pattern of the head make the animal quite difficult to see, but this is no argument for calling the head cryptic. However, on one occasion I was watching a sow with her two small cubs which were playing near the sett which was in a slight hollow. Suddenly a dog barked loudly, and although out of sight it could not have been far away. The cubs reacted by going below ground at once, but the sow which was standing at the time flattened herself on the ground at the entrance and partly covered her head with her fore paws. The camouflage was perfect. This type of behaviour, more often seen when actually under attack, demonstrated very well how a badger can on occasions make itself inconspicuous.

It was R.I. Pocock (1911) who first suggested that black and white patterns of mammals could act as warning signals. He argued that animals which had these patterns were nocturnal or crepuscular, and usually had a very potent method of defence, such as the powerful bite of the badger or the nauseating, deterrent stink of the skunk. He also pointed out that these animals fed mainly on small prey which did not move fast, rather than on agile animals which would be forewarned by the obvious nature of the animal's colouring. He also noted that these animals also made little attempt to conceal themselves, but shuffled about noisily while foraging.

From my experience, I strongly endorse this theory as it applies to the badger. However, Maurice Burton (1957) argues that if this is warning coloration, what enemies has the badger to warn off? This is a good question as badgers today have very few enemies beside man. One could argue that the badger has been around a long time and an animal such as the wolf could at one time have been a possible aggressor. Foxes too have been known to attack badgers, although normally they keep out of each other's way unless disputing a refuge or defending young cubs. However, I would agree that these are not very convincing arguments in favour of the warning coloration hypothesis.

But what of the cubs? Surely it is they who need the warning colour. Cubs certainly have their enemies, particularly foxes, and foxes and badgers often live in the same burrow system. I consider it is most significant that cubs acquire the same pattern as the adults almost from birth; because by this means they are able to share to some extent the advantage that the adults enjoy. A fox, having once had a painful encounter with an adult badger would associate this with the conspicuous facial pattern and might well think twice before attacking a cub showing the same markings. The situation is comparable to the advantage gained by some insect mimics, which although harmless themselves, resemble those which have stings or a noxious taste.

I believe this hypothesis is also supported by a cub's reaction to attack. Thus, it at once turns to face the aggressor, showing the facial markings to the best advantage and at the same time fluffs up its coat, making it look much larger than it really is. It also makes a series of most menacing noises. I realised just how effective this behaviour was when watching a cub digging for pig-nuts quite oblivious of my presence. On coming to within a metre of where I was standing it suddenly caught my scent, made an explosive snorting sound and erected its fur like a bottle brush as it faced me. Some of the fur seen from in front looked like a large circular ruff surrounding the head as the pale underfur showed up conspicuously. It certainly succeeded in making itself *appear* formidable.

The pattern may also serve as a useful warning to other badgers to keep their distance when feeding. Although badgers usually forage independently, when the cubs are young they often keep within a short distance of one another. However, if one finds a sizeable item of food such as a dead bird and another approaches, it will at once display its pattern by facing the trespasser and make a warning sound.

There is little doubt that the pattern also serves as a recognition signal. A badger's eyesight is not good, but the white pattern is more effective, even in poor light and easily distinguished by other badgers. David Humphries wrote, that if he held a badger's mask in front of his face, the badgers appeared to accept him as one of them if he crawled about on the ground nearby, but down wind. My own experiments using a stuffed badger also confirmed that other badgers at once recognised the pattern. They also reacted to their own image in a mirror as if they recognised it as a badger.

So, to understand the general coloration of the badger it is best to consider it in terms of its reactions to possible aggression. It has three alternatives. First, it can turn and face the aggressor; it then displays its full warning colours. Second, rather unusually, it can flatten itself and blend as well as possible with its background hiding its face between its paws. Third, it can run away; in which case its head, held low, would not be visible to the aggressor and the body would be inconspicuous because of its camouflage.

Variation in Hair Colour

In many mammals hair colour is very variable and its genetic control complex. In the badger, the various colour forms which occasionally occur within the population have not been worked out genetically with any certainty, but comparable conditions occur in other species. There are at least three easily recognised colour variations: melanistic, albino and erythristic; but there are also intermediates between these three categories caused by varying amounts of melanin deposited in the guard hairs. One of these may be described as a semi-albino. There are also colour varieties which do not fit into any of the above categories, notably a sandy yellow badger.

I know of no badger which is totally melanistic, but I have seen very black animals where the concentration of melanin in the hairs was unusually high, although the white parts were normal. P.J. Prosser also wrote to me about a 'jet black badger' which he watched near Dursley in Gloucestershire.

True albinos are seldom pure white, although when very young they approach this condition. Usually they are more of a cream or biscuity colour. Much also depends on the colour of the local soil as the hairs are liable to get stained. Albinism is due to the almost complete absence of melanin, so the facial stripes are usually invisible, although in some cases you can just make them out. The hair on the underside of the body should theoretically be as white as that on the back and sides, but the ventral hairs are more liable to get stained and so often appear slightly darker. The claws and tip of the snout also lack pigment, and the absence of melanin in the cells behind the retina causes the eyes to appear pink.

Albinism is a recessive factor, so when an albino mates with a pure breeding, normally-coloured animal the cubs will all be normal. However, these cubs will all carry the recessive gene for albinism, and if such animals mate the chances of an albino in the litter will be one in four in the usual Mendelian manner. With a rather limited dispersal range and some inbreeding a higher proportion of animals carrying this recessive gene will occur in some populations, hence albinism tends to be a fairly local phenomenon. Dorset, Kent, Berkshire and Essex are counties where albinism occurs regularly.

In some regions, accounts of white badgers are almost legendary and certain ancient setts are associated with their periodic appearances. One such area is near Dorchester, where albinos continue to be recorded. For example, one adult was killed there by a car in February 1962 and the skin sent to Dorchester County Museum; and later that year a young albino sow was found dead in the same locality. Then, three years later, D.H.B. Brown photographed a very large albino boar in the same area; and later that year saw a family of cubs at a place a few kilometres away, one of which was an albino. Then in 1979 a local farmer disturbed some diggers who were about to shoot an albino they had just dug out and in 1980 he also found two dead albino cubs at the same sett: they too had been shot. On the same farm another albino was seen in 1983/4 and I watched there myself. The site was difficult for watching due to much vegetation so the entrance was obscured, but this made the appearance of a large albino all the more dramatic. One moment there was nothing, the next, there it was, large, conspicuous, pale biscuit coloured and almost unreal staring in my direction with pink beady eyes. It was a strange and memorable experience. Other albinos have been seen near Weymouth in the same county.

R. Beaumont also reported albinos in Berkshire. At one sett he was watching there were three albinos, two of which he was able to photograph.

One was an adult and the other two were younger animals presumed to be the offspring of the first, but from litters of different years.

The most complete account of any albino badger was given by G. Burness (1970a). It represented a wonderful piece of sustained field work over more than nine years, with the help of two enthusiastic badger watchers, Gary and Philip Cliffe. This albino boar was seen on more than 200 occasions and its life history documented.

The badger, aptly named 'Snowball', was born as one of a litter of three in 1961. The other two cubs were normally coloured. The albino cub was treated in the same way as the others although in play there were slight signs of persecution. As Burness wrote, 'he always seemed to be in the front of every chase'. By the autumn he became rather less sociable emerging at a different time from the other cubs and going off to forage on his own. He remained in the parental sett for his first winter, which was a very severe one, and did not leave it until the following June when a new family of cubs was much in evidence at a sett in the same hedgerow. He could not be found for a month or so and at one time was thought to have died, because an albino boar was reported killed by a car not far away. However, this turned out to be an older animal and 'Snowball' was eventually located at a sett about 400 m away from his original home. He was living amicably with a large, very old boar and a small sow, both of usual colour.

During his second winter he became mature and was seen to mate with the small sow in February. Four cubs were born to this sow the following year, all of normal colour, and in succeeding years when she had litters, the cubs were always normal. 'Snowball' was seen quite often at the same sett with her, but in later years less frequently. He was last seen by Burness in late 1971.

There is some confusion over the colour of so-called erythristic badgers, due I believe to the fact that semi-albinos have also been described as such. The true erythristic form has a distinctive gingery appearance on back and sides, while the normally black underparts are a sandy red. The colour appears to be due to a different form of melanin in the guard hairs. The density of the pigment and its degree of reddishness varies slightly from one animal to another. There is some doubt about the colour of their eyes. The majority appear to be light brown, but some with reddish eyes have been described. Erythristics have been recorded from many counties, but like albinos they are more common in certain localities where, presumably, the relevant genes are more frequent in the population.

Semi-albinos are somewhat intermediate between erythristics and albinos. These are badgers in which pigmentation is greatly reduced so they appear a very pale yellowish brown. Unlike the true albinos the facial stripes can easily be seen, but the eyes are pink and the nose pale brown.

In Essex 'light-coloured' badgers have been seen in various parts of the county. W.W. Page told me of one very large sett which was of particular interest as it contained both normal and light-coloured badgers. One semi-albino boar was seen there many times during the period 1964–6. It was very pale with a yellowish tinge, but having quite obvious facial stripes and rather darker hairs on the ventral side of the body and legs. It was of particular interest that in 1963 a true albino cub was born at the same sett and was presumed to have been sired by the semi-albino.

I take the view that the semi-albino as described above is a distinct genetic variety quite different from the erythristic form, but until work on

GENERAL CHARACTERISTICS

Figure 2.5 Semi-albino: note indication of darker colour in eye-stripe and body

the genetics of hair colour in badgers has been classified, it cannot be ruled out that the difference between them is merely one of degree of pigmentation.

The range of hair colouring in badgers it not confined to those described so far. I published an account of one badger in the Cotswolds (1948) as 'a bright sandy yellow, more yellow than a sandy cat'. This sow was seen in 1945. In the summer of 1957, D.A. Humphries watched many times in the same wood and saw 'a fairly yellow boar and a very yellow sow: they were seen several times in good light. The sow had two cubs which did not show any yellow colour.' The yellowness of these animals was due to the colour of the non-melanic part of the hairs, the dark area of each guard hair was normal.

It is important to distinguish these genetically coloured forms from local variations due to environmental conditions. For example, in the Brendon district of Somerset, all the underfur and the lighter areas of the guard hairs are very reddish in colour due to staining by the local red sandstone soil. In another area where badgers are living in an ironstone deposit they have been described as 'positively scarlet'. No doubt the Durham badgers that tunnelled into a coal tip could be described as pseudo-melanic!

The colour of the pelage also varies with the age of the animal. By the time the cubs appear above ground (if their colour is typical), they are silvery grey, although there may be some staining of the fur in some soils. By the autumn of their first year the hairs are usually very slightly yellower. When they acquire their adult winter coat in the autumn of their second year they appear a darker grey. With very old animals there is a tendency to become lighter again, due probably to the decrease in the proportional size of the dark band in individual guard hairs.

GENERAL CHARACTERISTICS

THE DIFFERENCE BETWEEN THE SEXES

There are no differences in colour or pattern between the sexes so identification is often difficult under field conditions. However, there are certain other characteristics which, when taken together, can be helpful.

The adult male is more heavily built, and when seen head-on shows a distinctly broader head with fuller cheeks. The region between the ears is also slightly more domed. When seen from the side the nose appears blunter, the head shorter, and the neck thick. A much more variable, but nevertheless useful characteristic is the tail, which in many males is thinner and whiter.

The adult sow is a sleeker animal with a narrower head and neck, also her head is flatter and narrower between the ears. The tail is typically more tufty, and usually less white dorsally.

Figure 2.6 Adult sow

GENERAL CHARACTERISTICS

Although the width of the head is probably, in the field, the most useful indicator of sex, as R.W. Howard showed by skull measurements, only those with particularly broad or narrow heads are true to sex and there is a grey area of overlap where this criterion is unreliable.

Very occasionally you can see the genitalia of a badger under field conditions when it sits up and scratches itself. If you examine a dead animal the scrotal sacs of the adult male are evident, but in young animals they are not prominent. The penis is not visible as it lies under the skin in a forward projecting position and when in use it protrudes through an aperture in the skin.

One skeletal feature which is characteristic of most carnivores is the baculum, or os penis. This is a rod-like bone with a groove ventrally for the passage of the urethra; it has an expanded head and supports the penis. The

Figure 2.7 Bacula. Four examples of known age from Susan Fargher's work on material collected by Dr P. Morris. All views dorsal and lateral at life size

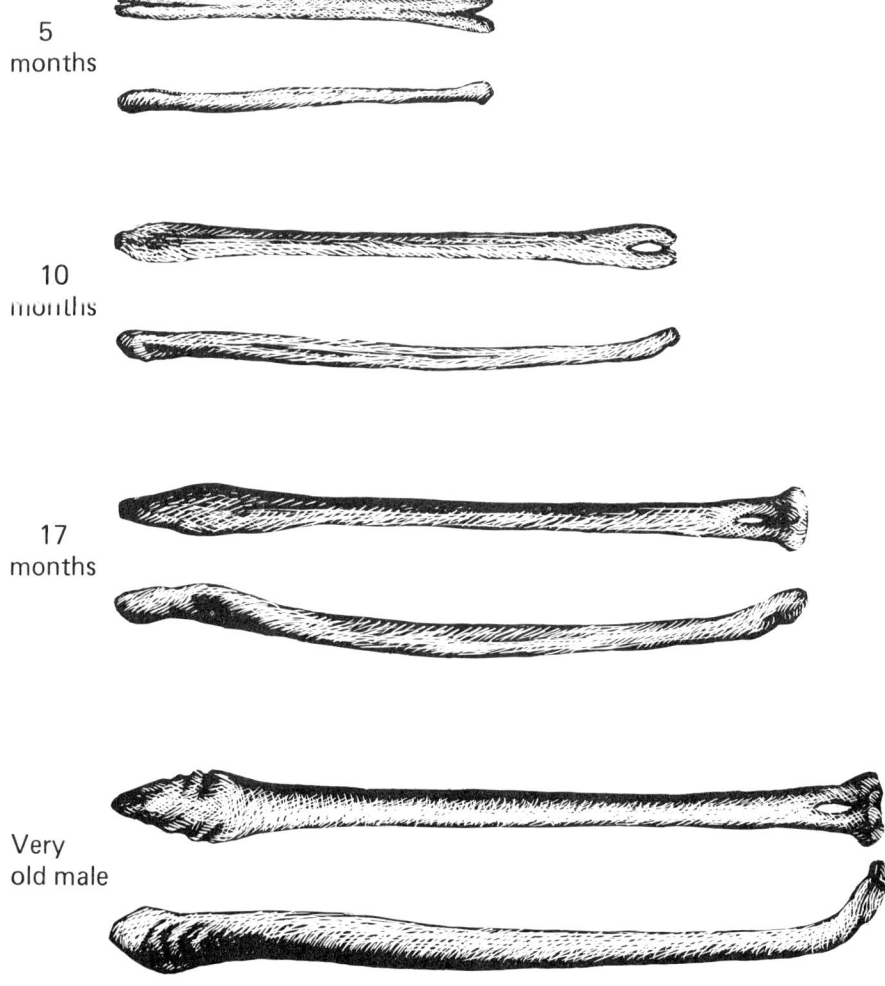

5 months

10 months

17 months

Very old male

GENERAL CHARACTERISTICS

baculum may be of some significance in inducing ovulation during copulation. Incidentally, in south-west Scotland the bacula of badgers are sometimes polished and used as brooches. In both sexes there are three pairs of nipples.

Immediately below the tail and above the anus in both sexes there is a large pouch, the sub-caudal gland which secretes a cream-coloured fatty substance with a faint musky smell. The function of this gland will be discussed later (p. 143).

THE SKELETON

A badger's skeleton endorses much of what has already been said about the fossorial adaptations of the species (p. 15). It has the general mustelid proportions, being low slung and rather elongated, but it gives the impression of great strength with its rather rigid spine and short, but very strong limb bones. The general arrangement may be deduced from Figure 2.8. For details of some of the individual bones the book by Lawrence and Brown, *Mammals of Britain, their Tracks, Trails and Signs* (1973), can be recommended.

The characteristic shape of a badger's skull is best understood by considering its main function. Primarily it has to house and protect the brain, accommodate the organs of special sense, eyes, ears and olfactory, and finally, deal with the mastication of food. The shape, size and strength of the skull is therefore a reflection of how a badger carries out these functions and the proportional importance of each.

The large brain case, small orbit placed rather far forward and the somewhat elongated nasal apparatus all reflect the relative importance of the structures they protect. But the median sagittal crest, the position and strength of the articulation of the lower jaw, and the comparative size of the zygoma are adaptations related to the tremendous strength of the bite and the type of dentition.

A badger's bite should never be underestimated as many a person can testify who has handled badgers, tame or wild. It used to be said that if caught in a spring trap, the marks of its teeth on the iron could sometimes be detected!

The closure of the jaw is brought about by three sets of muscles, temporalis, masseter and pterygoideus. In the badger, it is the temporalis which

Figure 2.8 Skeleton of a badger

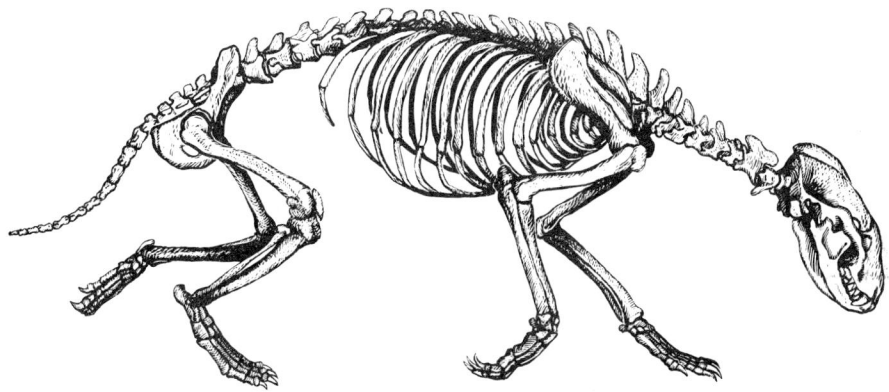

GENERAL CHARACTERISTICS

Figure 2.9 Adult boar badger skull: lateral and anterior views

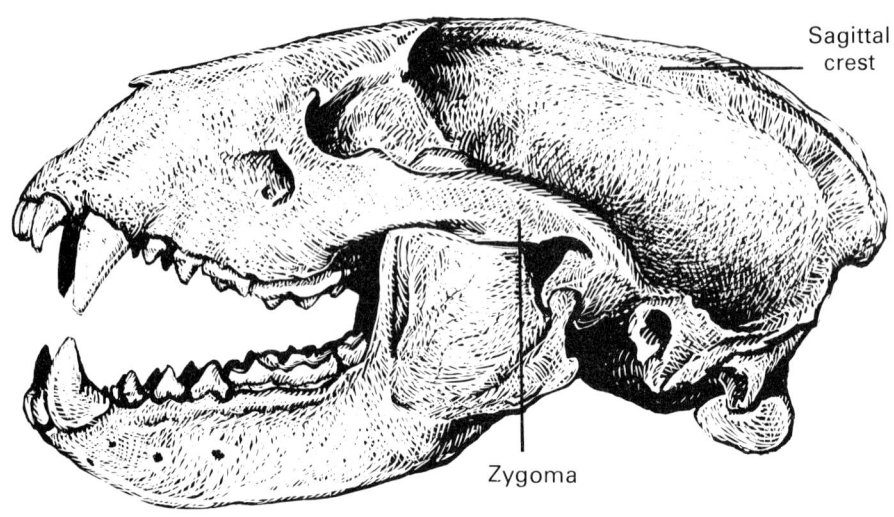

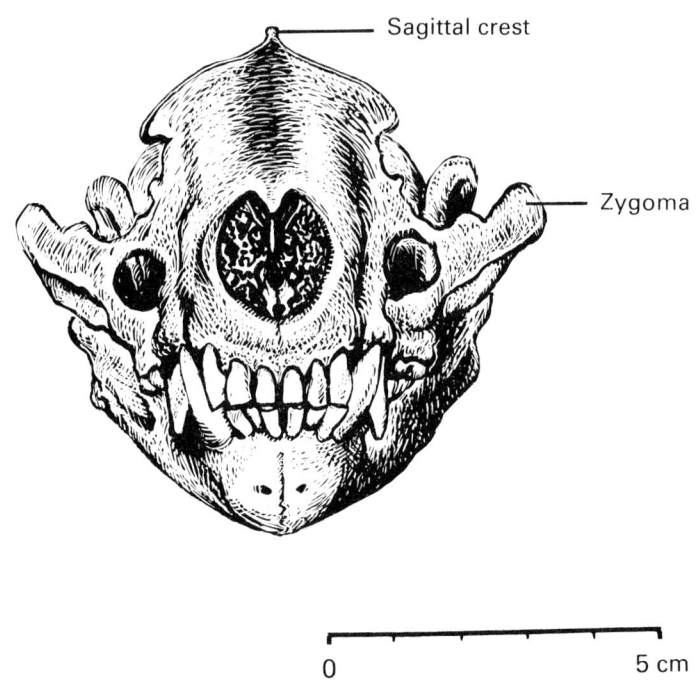

is extremely well developed and does most of the work. This muscle arises mainly from the lateral surfaces of the brain case and the fibres are inserted on the upper projecting portion of the mandible (see Figure 2.10). From the region of insertion the fibres fan out in dorsal and posterior directions giving great leverage to the jaw. To increase the surface area for the attachment of this great temporalis muscle the median sagittal crest has evolved. It forms the chief distinguishing feature of a badger's skull.

GENERAL CHARACTERISTICS

Figure 2.10 Side view of skull to show direction of main fibres of the temporalis muscle, which gives great leverage to the jaw. a=anterior fibres, p=posterior fibres. For further details, see main text

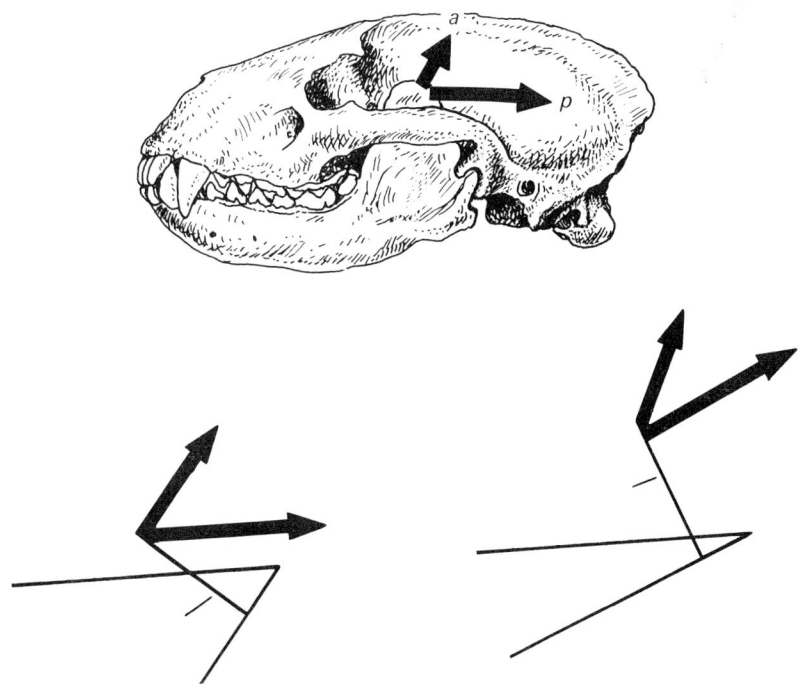

It is also noticeable that the point of articulation of the lower jaw is further forward than in other non-mustelid skulls. This gives more room for those temporalis muscle fibres which run backwards horizontally. These are the ones which are so important when the jaws are nearly closed and hence when the bite is strongest. Because of this great pressure exerted by the temporalis the joint with the mandible needs to be strengthened to prevent dislocation. This is achieved by enlargements of the bones of the socket both in front and behind to form a groove. The jaw fits into this so completely that with an adult badger's skull dislocation is impossible and only a fracture will allow the jaw to come away from the skull. It is possible, however, that this type of articulation may be as important for accuracy of bite, as for strength.

The juvenile skull has no sagittal crest. This appears at about ten months when the temporal ridges coalesce in the mid line. These temporal ridges appear in the young cub as lines on the skull surface marking the upper limit of the temporalis muscles on the sides of the brain case of the skull (Figure 2.11). As the skull grows and the muscles increase in size, these temporal lines gradually migrate nearer to the mid line, becoming more obvious as they form slight ridges. They eventually coalesce to form the sagittal crest. The gap between the temporal lines or ridges in young skulls is therefore a useful guide to the age of the skull (Figure 2.11).

After establishment of the median ridge, growth in depth of the crest is rapid for the next two or three years, but after that growth gradually slows down. The maximum depth recorded is 15 mm. Actual size, however, is an

GENERAL CHARACTERISTICS

Figure 2.11 Development of the skull up to one year, showing formation of sagittal crest. Top: 6 weeks, 8 weeks, 12 weeks, 16 weeks; bottom: 7 months, 9 months, 10 months, 1 year

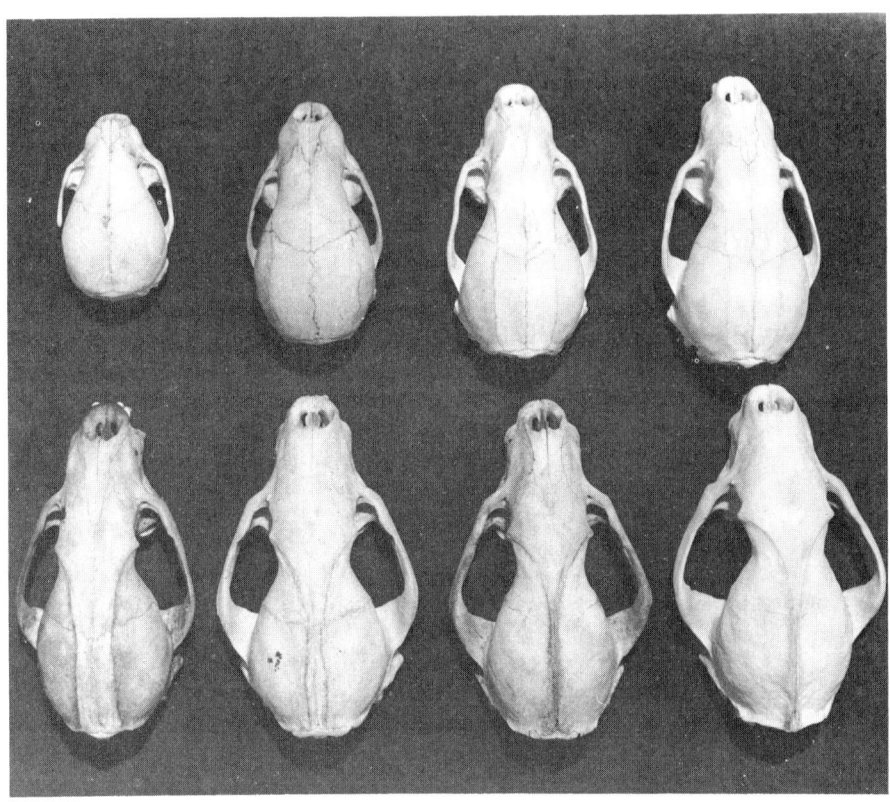

unreliable criterion of age as there is much individual variation. Skulls of one to two years can usually be recognised by the persistence of rough porous bone along the apical line of the crest which is indicative of continuing rapid growth. This subsequently becomes covered with smooth, hard bone. Also, the sides of the crest in younger skulls are smooth, but become progressively roughened by bone accretion in older animals. In old skulls, there is a lateral thickening of the crest posteriorly and the crest comes to overhang the back of the skull.

Further factors for estimating the age of a badger include the degree of closure of skull sutures, the time when the long bone epiphyses fuse, the characteristics of the baculum and the wear of the teeth. These are summarised in Appendix I.

Determining the sex from skull features is not easy, especially in older skulls. However, in younger animals the shape of the nasal bones appear to be significant. From Russian material, Ognev (1935) states that in the female the nasals are broader and shorter, having more rounded posterior regions, while in the male they are longer and narrower. This difference was confirmed by Hancox (1973) in a limited number of skulls from Britain.

GENERAL CHARACTERISTICS

DENTITION

The badger, although a member of the *Carnivora*, is omnivorous in its diet, and this is reflected in the dentition. Thus, the incisors, canines and front premolars are typical of a carnivore, but the last premolar and the molars are greatly modified for crushing and grinding, being broad, flat and multicuspid and thus more typical of a herbivore.

The dental formula for the permanent dentition is:

$$\text{Incisors } \frac{3}{3} \text{ Canines } \frac{1}{1} \text{ Premolars } \frac{4}{4} \text{ Molars } \frac{1}{2} = 38$$

However, the first premolar is often absent, and if present is always vestigial. Of 212 skulls examined by Hancox, 16 per cent had all four vestigial premolars present, 14 per cent had three, 61 per cent had two, 3 per cent had one and 6 per cent had none. Of the ones that had two, 57 per cent had them only in the lower jaws.

In Eastern races (in Russia and Siberia), there is usually a complete absence of this first premolar. In this respect they resemble the mink, polecat, stoat and weasel. The otter is intermediate, normally having four premolars in the upper jaw and three in the lower, all of them functional.

An unusual variation of the badger was described by Fullager *et al* (1960) in which the large upper molar was replaced by two separate and smaller teeth, each having three roots. As four vestigial premolars were present in addition this brought the full number up to 40.

The milk dentition of the badger is expressed by the dental formula:

$$\text{Incisors } \frac{3}{3} \text{ Canines } \frac{1}{1} \text{ Premolars } \frac{4}{4} = 32$$

but again the first premolar is not always present and if it does occur it is often shed very early or does not penetrate the gums. Similarly, the milk incisors although present, may not all penetrate; this is often the case for incisors 1 and 2. Their absence may well be looked upon as an adaptation for suckling.

The time of eruption of the milk teeth appears to be variable. But E. Overend gives the following sequence:

- 4 weeks   Milk canines erupted in both jaws.
- 4½ weeks   First signs of upper premolars.
- 5 weeks   First signs of lower premolars. Upper incisors 2 and 3 through, 1 did not penetrate. Lower incisors all absent or did not show.
- 6 weeks   All premolars through in both jaws.

The first permanent teeth, the upper incisors, erupted at ten weeks; and the lower incisors came through a week later.

The skulls of known age I have examined confirm that the first permanent incisors appear at about ten weeks in the upper jaw and the full transition to a permanent dentition occurs during the following six weeks. The sequence follows closely the order of teeth from the front backwards, those of the lower jaw lagging behind the upper by about a week in most cases. However, the

Figure 2.12 Dentition of cub aged 15 weeks, showing permanent teeth all present, with extra canine (c) and premolar (pm) of the milk dentition still not shed

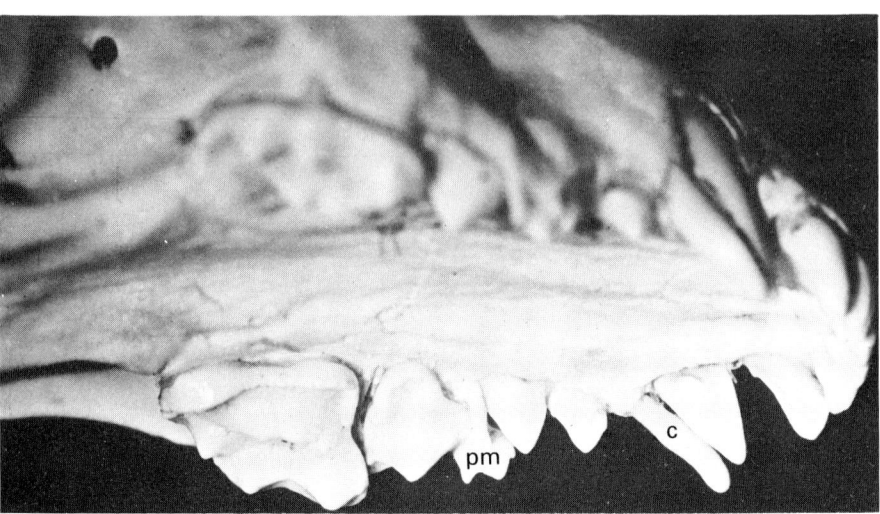

large carnassial (P^4) of the upper jaw is exceptional in erupting before the smaller P^3. This allows the former to become functional at an earlier age in relation to the large first molar of the lower jaw into which it fits when grinding. The detailed sequence is as follows:

Upper jaw I^1 I^2 I^3 C^1 (M^1) P^1 P^2 P^4 P^3
Lower jaw I^1 I^2 I^3 (M^1) C^1 P^1 P^2 (M^2) P^3 P^4

It is interesting that during the last phase of the transition from milk to permanent dentition around the 15th–16th week of life some of the milk teeth may still be present and functional in addition to the permanent teeth which have already come through. This is possible because some permanent teeth do not emerge through the same sockets as the milk precursors, but to one side. In a skull in my possession (Figure 2.12) there are four functional canines in the upper jaw and two in the lower, and a double set of P^3 in the upper jaw and P^4 in the lower.

Tooth wear has been used as a criterion for age, and in general terms this is useful, but it is not easy to work out a reliable picture of wear in relation to exact ages. Unfortunately, data from captive badgers are most unreliable as wear is abnormally light in such animals. In wild badgers, there is also variation in wear according to locality and presumably the food eaten, so the method must always be approximate. However, both Stubbe (1973) and Hancox (1973), using material from wild badgers, have described the general pattern of wear within wide limits, and this is certainly helpful especially for older skulls.

Hancox considers the upper molars, the lower first molars and the lower incisors are the best indicators of wear, the latter being especially useful for the 1–3 year group as during this period the molars show little sign of wear. The five stages of wear worked out by Hancox showing increasing amounts of exposed dentine are shown in Figure 2.13.

Figure 2.13 Tooth wear: stages in exposure of dentine. After Hancox (1973)

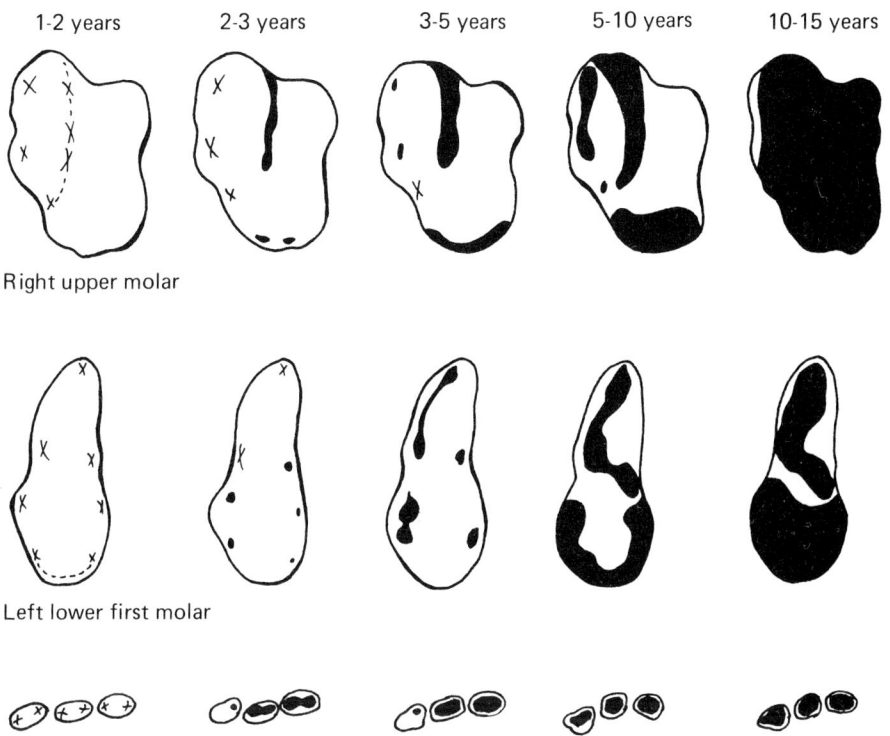

3 The badger's home and environs

SETTS

The burrow system of a badger is known as a sett. The term appears to be derived from the ancient word 'cete' which was the collective noun for a group of badgers. Today the word 'sett' refers only to its home.

Setts provide shelter for badgers during the day and are also used for breeding. They vary greatly in size, the extent of the tunnels and the number of entrances. If there were such a thing as a typical sett it would have 3–10 large entrances leading to an intricate system of interconnected tunnels and chambers. Outside each entrance there would be a large mound of soil excavated during digging operations.

A badger sett may be distinguished from a fox's earth by the much larger heaps of soil outside the entrances and the remnants of vegetation always present in the excavated soil. Occasionally some rabbit holes, if dug in loose soil, may cause doubt because of their relatively large size, but on closer inspection you find that within a short distance of the entrance the tunnels narrow considerably to no more than 150 mm if rabbits have made them.

The entrances of a badger sett are not less than 250 mm in diameter, are typically 300–350 mm and occasionally in old setts up to 600 mm. Newly excavated holes have rough edges but the walls of well-used entrances appear smooth and polished, due to the constant rubbing of the badgers' bodies.

The Spoil Heap

The mound of earth or spoil heap outside each entrance is sometimes very large and may contain 30–40 cu. m of soil weighing several tonnes. This may have been excavated by generations of badgers, although it is surprising how much can be dug out even in a single night. In a large sett, the whole configuration of the hillside may be altered by these immense digging operations. If the stratum in which the badgers are excavating is chalk, recent diggings may be seen from a great distance as white patches spilling down the otherwise green hillside.

Incorporated with the soil there is always some old bedding in the form of hay, bracken, leaves, etc. (p. 46) and by breaking up some of the lumps of soil you often find badger hairs.

Figure 3.1 Badger sett in elder thicket, showing spoil heap

In May, if the spoil heap outside a well-used entrance is flattened and hardened it is a good sign that cubs are present.

A broad furrow leading from the entrance across the spoil heap with loose earth all around is a sign of recent digging, the furrow being made by the badger when removing the soil.

The Underground Labyrinth

Local conditions differ so greatly that no two badger setts are alike, but even when setts are dug in similar soil and comparable situations the pattern of tunnels can be quite different. Each badger seems to be its own architect.

The simplest form of sett is one which has been dug as a temporary shelter. It consists of a single tunnel which usually bends 1 m or so from the entrance and ends in an enlarged chamber. A badger would not use such a sett for long without enlarging it considerably by making new tunnels and alternative entrances. The latter usually arise from side tunnels which run at a depth of about a metre and parallel to the ground surface; they often follow the same contour if the sett is on a slope and reach the surface again 3–4 m from the first entrance. From the first chamber, other tunnels are constructed which follow any kind of pattern according to the type of soil, position of rocks, or tree roots, and presumably the inclination of the diggers! More chambers are formed periodically and the whole labyrinth enlarges from year to year if the terrain is suitable.

In sandy soil, the tunnels may penetrate deeply into a hillside, and even when the site is flat they may reach a considerable depth. An abortive badger dig by Sir Alfred Pease (1898) towards the end of the last century makes the point very well. The sett, which was in a flat field, had only three entrances, but the badgers had been there for generations. A trench was dug about 2 m in depth, but the sounds of the dogs they had sent into the sett were heard to come from below, so they dug a further shaft from the floor of this trench to a depth of another metre. By this time they had cut through badger tunnels at three different levels. On hearing the badger and dogs still below them they gave up the dig in despair. So in this sett there were definitely four storeys of tunnels going down a minimum of 4 m.

In soils which are more difficult to penetrate because of the underlying rock, the labyrinth is often roughly at one level and may not penetrate deeper than a metre. This is a common feature in some Cotswold setts where the underlying limestone is particularly hard and the soil shallow; here they utilise the subsoil for their tunnels. Under these circumstances some tunnels may come too near the surface and the roof fall in. New entrances are often formed in this way.

Shallow setts also occur in the shales of the Quantock Hills in Somerset and in some chalky districts. A characteristic pattern here is for the main tunnel to go in less than a metre before turning at right angles along the countour for 4–5 m before enlarging to form the first chamber. However, in many setts dug into chalk the tunnels may be very extensive.

Badger tunnels are usually wider than they are high, so that in section they appear almost semi-circular. In most tunnels the height is usually about 250 mm and the width about 300 mm, but the dimensions can be much larger near entrances and in regions which act as passing places. In one very ancient sett dug in red sandstone, I was able to measure the dimensions of one of the main tunnels as badger diggers had made an abortive onslaught on it a few days previously. At a distance of 4 m from the original entrance, the tunnel was still 600 mm in diameter and a man (if he so wished!) could easily have crawled down it for some distance further. The

Figure 3.2 Plan of a large sett in Gloucestershire, covering an area of 35 ×15 m. The 94 tunnels totalled 310 m in length and had an average depth of 0.76 m (greatest depth 1.92 m). The volume of the burrow system was 15.28 m³, equivalent to 25 tonnes of soil excavated. By courtesy of the Ministry of Agriculture, Fisheries and Food

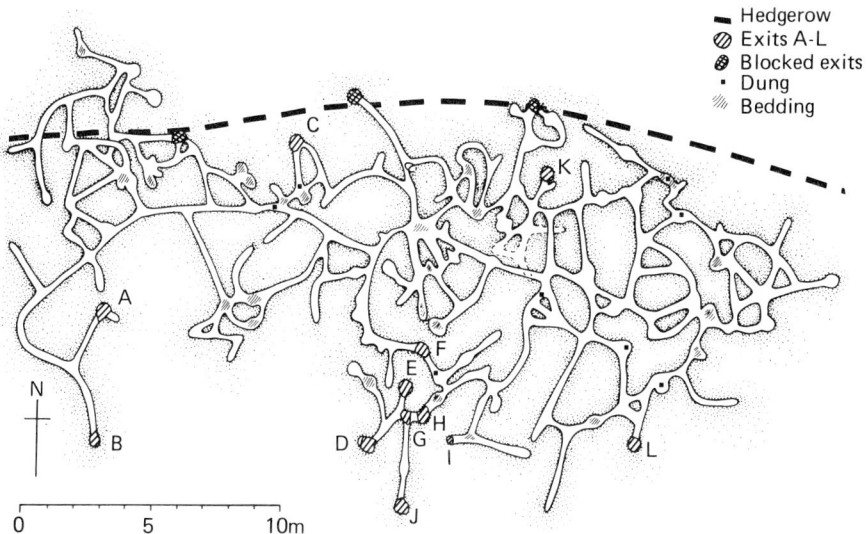

floors of the main tunnels are often compacted like concrete, due to the constant traffic. I have many times heard from above ground the pounding of feet on this hardened surface when a badger has been frightened and rushed down its sett. As you hear this pounding getting fainter and fainter it makes you realise how extensive some of these tunnels are. One in solid sandstone went in over 40 m (Wijngaarden and Peppel 1964) and in soft sand they may go in much further.

The chambers can be quite large. I have known those used for cubbing to be 600–900 mm in diameter and up to 600 mm in height, but more often the height is about 450 mm. The first chamber excavated usually develops eventually into a 'roundabout' for several radiating tunnels which may lead to further chambers, and is no longer used for breeding or sleeping. Humphries (1958) describes a sett in Gloucestershire where a large hollow tree stump appeared to be a major junction of underground passages—eleven tunnels radiating from it at three different levels. Chambers used for breeding or sleeping are often terminal enlargements of side tunnels, but they are not always cul-de-sacs.

Breeding chambers in well established setts appear to be 'well chosen'. Some are situated directly below large boulders or slabs of rock and in setts dug into uniform sand the roots of trees are often utilised in a similar way and so strengthen the roof. Nearly always the tunnel slopes upwards towards such a chamber, which helps drainage—a very important feature if young are to be raised successfully. Sometimes this incline takes the form of a kind of step. This gives a badger considerable positional advantage if attacked by a dog, as the badger lies above the step and is in an almost impregnable position.

Both sleeping and breeding chambers are filled with bedding and are big enough for more than one badger to sleep there. Judging from observations on badgers kept in captivity, it is likely that two adults normally sleep lying stretched out side by side and head to tail.

Setts need to be well aerated. This is usually achieved by having a number of entrances, some at different levels; this allows a flow of air through the system. This flow is well marked on cold frosty mornings when some entrances 'steam'. This happens when the moisture in the air, warmed by the badgers' bodies, condenses on reaching the colder atmosphere. It is a good indication of the proximity of a breeding or sleeping chamber to a particular entrance.

Ventilation holes also exist in some setts. Whether they are made deliberately or are utilised by the badgers after being formed accidentally is difficult to determine, but the latter explanation seems to me more likely. They are usually 30 or 40 mm across and lead from the ground surface to a badger tunnel below. Whatever their origin they presumably help the circulation of air in some setts. It is important when watching badgers not to choose a position close to one of these holes as a badger may scent you before emerging. It is disconcerting to hear a sudden snort from below as the badger takes fright.

Size of Setts

To the observer above ground, the size of a sett is not easy to determine, although the amount of earth excavated from all the entrances is often the most useful indication of the extent of the tunnels. The number of entrances is an unreliable guide to size as this is determined by a number of local factors of which soil type and human interference are probably the most important. For example, it is common hunting practice to block all holes before a hunt to prevent a fox using the sett as a refuge. This may alter the number of entrances as the badgers do not always react to this disturbance by quickly unblocking them again. Sometimes some holes remain blocked for months or even years.

The largest setts are those which have been long established and are in places where the soil is particularly suitable for digging. Some of these can almost be described as badger cities, they are so large. One sett in the Brendons, Somerset which I visited in May covered an area of about 1,600 m^2. Over 80 entrances were concentrated in this favourable situation of which 48 had recently been cleaned out. It had been undisturbed for years. However, sett size and the number of entrances bears little relationship to the number of badgers living there.

Badgers have an instinctive urge to dig even if the sett is already far too large for their needs, so some parts become derelict or unoccupied as the badgers tunnel elsewhere and excavate new chambers. Subsequently, the badgers may clear out old entrances not used for a year or more and renovate previous sleeping or breeding chambers. So a large sett with many entrances may alter its centres of activity to some extent from season to season.

I have known setts 'move' quite considerably in the course of 20–30 years. One on a hillside in Somerset started as a small sett with three or four entrances at one level. The position of these can still be made out by the old mounds although the entrances have become filled in over the years. Later, other entrances below them were excavated and later still others below that, so the active position of the sett has changed location by 20–25 m.

In a similar manner, setts can 'move' along a hillside following a soil stratum suitable for digging. This tendency may cause setts to approach each other to form an irregular line of entrances stretching for 1 km or more, so that it is no longer easy to distinguish one sett from another. The growth of these large setts and the problem of territory that results will be discussed in Chapter 9.

Ancient Setts

Badgers are very conservative regarding their homes. If generations of badgers have found certain setts suitable they will tend to go on using them unless new factors make this impossible. I have known setts where the badgers have been subjected at one time or another to digging, snaring and gassing but which are still occupied today. This, of course, is largely due to recolonisation by badgers from the surrounding, high density area, but it illustrates the importance to badgers of well proven sites. For this reason if you talk to certain countrymen whose memories go back 60 or more years, they will tell you where the badger setts in the neighbourhood were when they were boys and on investigation you find that most of these setts are still there today. There is no doubt that many setts go back several hundreds of years.

One particular sett at Ashlyns in Hertfordshire is famous both for its antiquity and its size. T.V. Roberts (1893) tells how the sett was certainly extant towards the end of the eighteenth century and was probably very much older than that. It was situated in a wide pit in the chalk with beech trees growing in and around it. In 1890, a great effort was made to eliminate this sett in order to get rid of mange in the foxes which lived there too. This was a time of great unemployment, so eight men were given the job. They dug steadily for ten days without coming up with the badgers! Michael Clark tells me this sett is still used by badgers.

The Mendip Hills are honeycombed with caves in the limestone and badgers are plentiful in the area. One ancient cave there known as Badgers' Hole was being investigated for archaeological remains by W. Balch. When digging out an area where the roof had fallen in many centuries before he found a number of badger bones along with those of elk and cave bear later estimated at 60,000 years old. Badgers still live in part of that cave today! In fact, Balch met one face to face in one of the tunnels. He also discovered in this cave an ancient runway he believed had been made by badgers.

The Origin of Setts

When new setts are formed, they are usually dug on existing territories as outliers of the main breeding sett. At first, they are only used spasmodically as alternative accommodation, but if one of these outliers is found to be suitable it may well be enlarged and eventually become another breeding sett. This often happens in some communities, the badgers moving from one to another according to circumstances. It also allows sows of the same social group to have their families on their own during the crucial period when the cubs are young.

Some setts are excavated where no previous digging has taken place, but it is much more usual for badgers to enlarge rabbit burrows. The latter are often dug in soils and situations which are also suitable for badgers, so they make a useful starting point. In Russia, Novikov (1956) mentions that they may also take over marmot burrows.

DIGGING

I have always liked Nicholas Cox's description of how badgers dig their sett. In 1721 he described the process as follows:

> One badger falleth on his back, another layeth earth on his belly, so taking the hinder feet in his mouth draweth the belly-laden badger out of the hole and having disburthened himself re-enters and doth the like until all be finished.

I think badgers must have changed their habits during the 250 years since those words were written! However, it is not easy to see exactly what happens during a digging operation except perhaps for the final stage when the badger emerges tail first from the entrance.

When digging, a badger appears to be very preoccupied with the process. I have on several occasions been able to stand close by an entrance and been showered with earth as the badger has backed out of the hole and kicked away the loose soil with its hind limbs.

When enlarging a tunnel, the badger makes rapid movements of its front limbs and claws to loosen the earth. When sufficient loose earth has collected, the badger arches its back and brings its hind legs forwards to push the soil backwards. In this way a heap of loose soil collects behind it. The badger then moves backwards, partly using its hind quarters as a ramrod and partly hugging the earth between its fore legs and chin. It then moves in

Figure 3.3 Digging: pushing back soil over edge of sett spoil heap and clearing back tunnel debris

a series of jerks until clear of the sett entrance. When the spoil heap outside the sett is large it will continue to drag the soil to the edge and then kick it away with its hind legs. Consequently, in sandy soil, a well marked furrow is often worn in the soft earth between the entrance and the edge of the heap. If the sett is in soft sand, before entering for a further bout the badger shakes itself vigorously to rid its fur of the loose sand.

When digging in very stony soil or in chalk or limestone, very large stones are removed along with the earth. Much effort is involved in winkling out these rocks when firmly embedded, and in chalk especially, you can often see the evidence of this in the deep furrows made by the front claws on the surface of excavated pieces. It is surprising what weights they can cope with: stones of up to 4 kg have been recorded. Sometimes, if the stones are very large, they may be pushed out of the entrance with the help of the snout, or dragged out backwards with the fore legs, and occasionally they may be carried in the mouth.

Paget and Middleton (1974a) described how they watched a badger dragging boulders of some 150 mm in diameter from its sett high upon a hillside, and 'each time the animal sent one rattling down the steep slope to the stream below, it stopped and looked down the slope until the noise abated before continuing!'

On some nights, digging may continue for two hours or more and huge quantities of sand may be removed. Usually, one animal does the digging, but occasionally others join in. Donald Bradnam described an occasion in May, when a badger emerged in good daylight and started digging. It was soon joined by another—this one already covered in sand. Shortly afterwards a third badger came along and helped in the digging and one after the other they brought out sand from the hole. Bradnam timed the operation and found that a load of earth was removed at exactly half minute intervals by these three badgers.

Outside one sett on the Quantock Hills in Somerset, which was dug into clay with some shale, I found in several consecutive years a few clay balls near one of the entrances. These rounded structures ranged from golf to tennis ball size. When sectioned they showed concentric rings indicating that they had been formed by some rolling action and incorporated in the clay were numerous badger hairs. I have since found rather similar structures made of red clay outside setts in the Brendon Hills. Beatrice Gillam has also found them in Wiltshire made of chalky clay. It is difficult to be quite sure how badgers make them, although Michael Clark actually photographed a badger rolling one between its front feet as it backed out of the entrance. He made the suggestion that they might be made when a badger's paws were coated with sticky clay and it tried to clean itself below ground; it would then back out, rolling the clay into a ball with its fore paws as it did so. However, Göranssen sent me a transparency of a Swedish badger showing a number of clay balls sticking to its fur and he believes that when clay in the tunnel has exactly the right stickiness it adheres to the fur in patches and as the badger moves along the tunnel the patches enlarge with further accretions and rotate on the fur whenever contact is made with the clay of the tunnel so forming concentric rings with hairs incorporated. It certainly seems to be true that these clay balls are only found at particular setts where the clay is very sticky.

BEDDING COLLECTING

Bedding collecting is one of a badger's most characteristic activities. The bedding is gathered up in bundles and brought back for storage in the sleeping and breeding chambers. The material used varies according to availability, the badgers of each sett tending to use the same type of bedding each year according to the season, although they are not averse to change if unusual opportunities occur.

In one sett in hill country near Taunton, Somerset, where there are no arable fields, the pattern is a very regular one. In the summer and autumn they use grass, but after the bracken has died and become dry they use this until April. In May they switch to bluebell leaves which are brought in moist and green.

In late summer there seems to be a general tendency over the whole country to use rank grass or hay whenever this is available, even when good alternatives exist. For example, the badgers from one sett in a copse on a steep slope which separates two fields had a choice between hay from the lower field and straw from the upper. They invariably chose the hay, although it was further away and much more difficult to bring back up the steep slope. If no long grass is available they will certainly use straw. Glennis Walton told me how some badgers in the Mendip area of Somerset brought back a complete bale of straw. Unfortunately for the badgers, this unusually large bundle got stuck in a fence just short of the sett, but undeterred they pulled it to pieces and eventually brought all the straw in. Straw appears to be particularly welcome in the winter in places where bracken is not available. I have known badgers collect it from a field, where it was put out for cattle during hard weather, and laboriously bring it back more than 80 m through difficult terrain to their sett in a wood.

In deciduous woodland, the setts are often situated near an edge where the wood butts on to grassland; this may well be significant, not only in terms of a badger's feeding ecology but also because it facilitates the collecting of bedding. However, if the sett is some distance from fields some badgers make use of dry leaves during the winter; oak and sweet chestnut are particularly favoured.

Other substances used during the winter include the fronds of male fern, the dry trailing branches of wild clematis when these are within reach, rock rose and even gorse and thistles!

Figure 3.4 Bedding gathering. Left to right: use of chin to bundle bedding; dragging backwards; and down into the sett

In May, when the cubs are larger, but still not weaned, some form of green material is used, especially bluebell leaves and dog's mercury. Susan Paton told me that in North Wales, daffodil leaves and flowers were also occasionally taken down, and in damp woodland in the west of England, the fronds of hart's-tongue ferns are regularly used during this month.

The method used to collect bedding varies according to the material. When gathering grass they scratch it up with the help of the front claws which act like hay rakes, but when it is tough they get a grip with their teeth, arch their neck and shoulders and pull with great strength, making much noise in the process.

Tough material such as bracken is usually bitten through before bundling and the same applies to male fern fronds, but I have known male ferns pulled up by the roots and the whole plant carried back. The fronds were then bitten off near the sett, and the old rhizomes left scattered about.

When leaves are used they scrape them together with their fore feet to form small heaps before transporting them.

Eunice Overend told me how she used to give newspaper as substitute bedding to a cub she was rearing. It would never bring the paper back whole, but would always shred it first, then bundle the pieces together and drag them backwards to its day nest in the bend of the stairs. It left a trail behind which looked like a paper chase!

Whatever the material, the normal method of bringing it back to the sett is to make it into a large bundle which is hugged to the chest, using the chin and fore legs to keep it in place. The badger then proceeds to shuffle backwards in a series of jerky movements of the hind limbs, more or less sliding on its fore limbs. This backward movement is surprisingly fast, but periodically the badger stops to listen or to retrieve any bedding that has become loose from the bundle. The badger's direction is seldom at fault although it's not looking where it is going, and it disappears into the sett entrance backwards with remarkable precision.

Occasionally, unorthodox methods are adopted. John Webster told me how he watched a badger in Cumbria bringing back bedding in the usual way, but because the last slope was very steep, it picked up the bundle in its mouth and took it into the sett. Eileen Soper (1957) also recounted how a sow, on emerging from a sett, found a bundle of bedding outside and took it up in her mouth; she then walked towards the entrance, but after a few paces turned round and brought it in backwards in the usual manner.

The bedding is collected in quite a systematic way from a particular area. On one occasion, in October, a sow collected a bundle of beech leaves from about 3 m from where I was standing in the low fork of a tree, and on the next trip it came still nearer without suspecting my presence. When it came once more, I anticipated that it would come right up to my tree and sure enough, it collected more leaves from within a few centimetres of where I was standing! In some places where the setts are surrounded by bracken, the whole area becomes progressively denuded of dry fronds as the winter proceeds.

The route used by the badgers when bringing back bedding is often very difficult to negotiate. At one sett the entrance was at the foot of an almost vertical bank some 3 m high and the best hay was above the bank. I could hear the badger shuffling back with its load from my position on a level with the opening, but could not see it until it reached the top of the bank. However, on reaching the edge it did not hesitate for a moment and slithered down the bank backwards, grasping the bundle as closely as possible.

The amount of bedding carried each time can be quite large. J. O'Connor found that some bundles of moist grass he investigated weighed as much as 0.5 kg. On some nights badgers may take in a few bundles and then go off to forage in the usual way, but occasionally they really have gala nights and a single badger may bring in 20–30 bundles one after the other. Sometimes one animal which is bringing in bedding appears to stimulate others to do the same and in April I saw three adults doing this at the same time, a boar and a sow bringing it back to one entrance and another sow to a different one.

Eileen Soper (1955) described how she watched a boar and sow both hard at work bringing back bedding in early June:

> The sow came with her bundle, a little cub trotting in front; the bedding was taken in, followed by the playful cub. The boar also took part in the work which was now started in earnest. The adults came up the path repeatedly with their loads, the cubs running beside them. They were bubbling over with excitement, squeaking and yelping as they played; getting in the way of the parents; running on ahead; returning to them, then back to the sett as though the whole business was proceeding at much too slow a pace for their liking.

The bringing in of bedding seems to be an inherited action, as it took place in cubs brought up in captivity from the time they were only two or three weeks old. At 8–9 weeks they collected small piles of grass or leaves and started to move them, but then they appeared to forget and did something quite different, but by 12 weeks they carried out the operation in a much more effective manner.

Factors Affecting Digging and Bedding Collecting

Digging and bedding collecting are often associated. The usual pattern is for part of the sett to be dug out first and then bedding is brought back. For this reason these two activities will be discussed together and from a seasonal standpoint.

Digging and bedding collection can take place in any month of the year, although there are certainly peaks of activity which are correlated with important events in the lives of the badgers. The most unpredictable diggings occur when an entrance which has not been used for some time is cleared of leaf debris and earth and subsequently used by one or more of the sett's occupants; this can occur during any month.

Keith Neal and Roger Avery (1956) carried out observations on a single, large, permanently occupied sett near Taunton, Somerset over a period of four years, involving 800 observations of which more than half were night watches. Digging and bedding collecting activities were noted and from these data I have attempted to correlate their frequency with such factors as the birth of cubs, the number of litters, and fluctuations in the number of badgers at the sett. Bouts of digging were far less frequent than bedding collecting, but both activities showed peaks which varied only slightly over the four years. The results are shown in Figures 3.5–3.7.

It is convenient to start with the period June–July, as this is the quietest period for digging and bedding collecting in south-west England. This is the time when the cubs are becoming independent and the whole social group is more concerned with feeding. At this particular sett there was hardly any digging or bedding collection in June or July except for the occasional entrance being cleaned out in the latter month. However, I have noticed in other localities where the family has moved to an alternative sett in July that this is often associated with considerable digging activities, often at both setts involved.

Sometime during the period August–October, a lot of digging took place each year, but the peak time was variable. Two factors appeared to influence the timing. The first was the weather; for example, there seemed to be far less digging during periods of drought. This could have been due to the hardness of the soil at that time, but I suspect that another important reason was the longer time needed to find food during very dry periods, which left less time for digging. The second factor appeared to be population numbers. For instance, few badgers were present in August and September 1955 compared with other years, but when more badgers arrived at the end of September a great deal of digging took place the following week. Digging during the August–October period was followed by the taking in of much bedding and presumably both activities were connected with preparations for the winter. Huge amounts of earth were excavated at this time and it is logical to assume that this is the main period of the year when new tunnels and chambers are dug out in a permanent sett.

Very little digging took place in November and the first half of December, which is the quietest period of the year in a badger's life, although small amounts of bedding were taken in during early November.

In the second half of December and in January, small-scale digging was renewed and during the latter month significant amounts of bedding were brought back. I am convinced that bedding collection at this time is connected with preparations for the birth of the cubs and takes place at any time when the weather is suitable within a few weeks of birth. In 1954 bedding was brought in on 2 January (the first collected since 20 October), more was collected on the 4th and 6th; the entrance was cleared on the 11th, and after three consecutive dry nights further bedding was brought in. A wet week followed, but on the 25th following another spell of dry nights, more was collected. The first litter of cubs was born at the beginning of February. As can be seen from Figure 3.6, a rather similar pattern occurred in succeeding years. So bedding collection at this season provides excellent

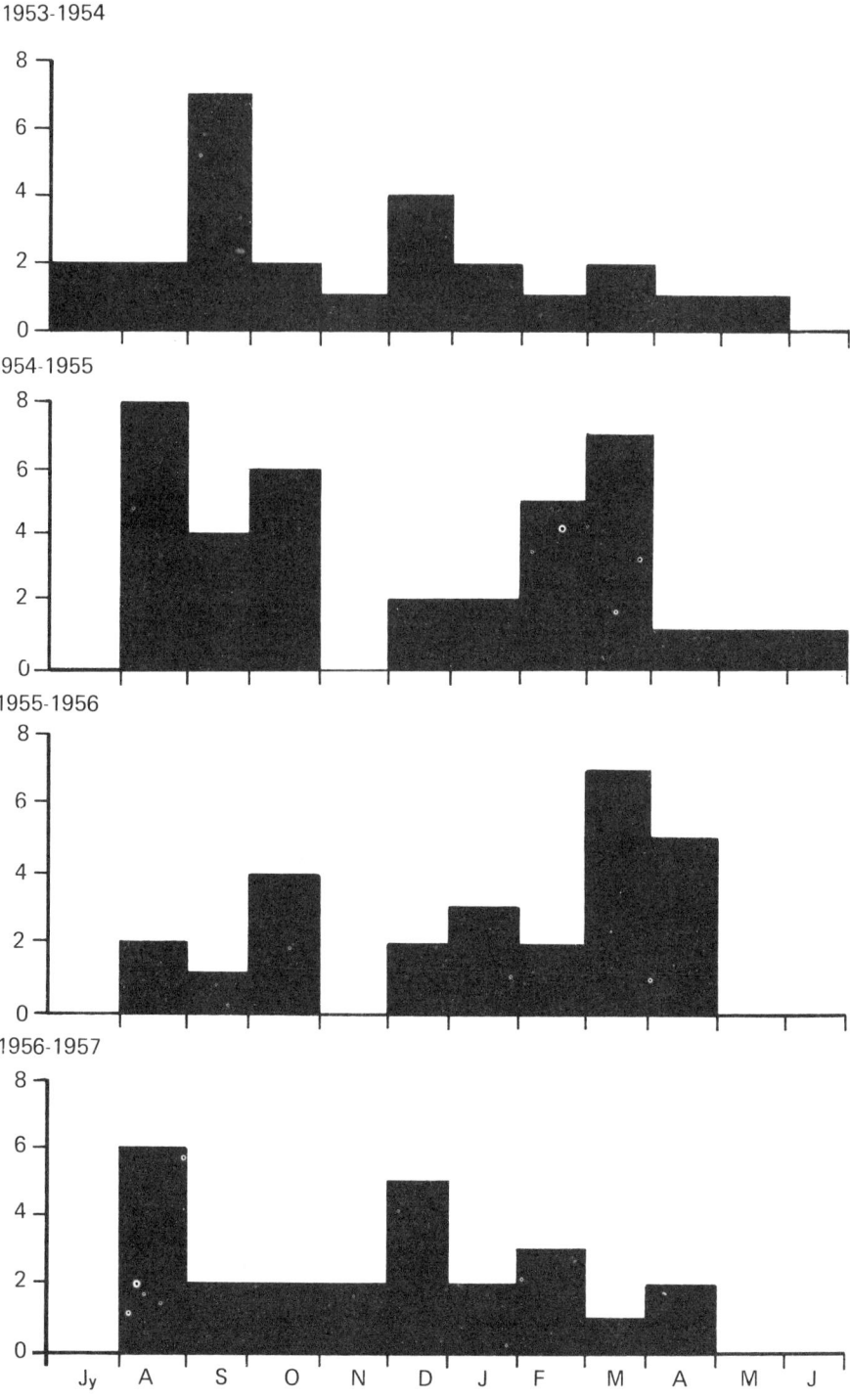

Figure 3.5 Number of nights per month when digging was known to have taken place at one permanently occupied sett near Taunton, Somerset. Data from 260 observations over four years by K.R.C. Neal and R.A. Avery

THE BADGER'S HOME AND ENVIRONS

Figure 3.6 Number of nights per month when bedding collecting was known to have taken place at the same sett as for Figure 3.5

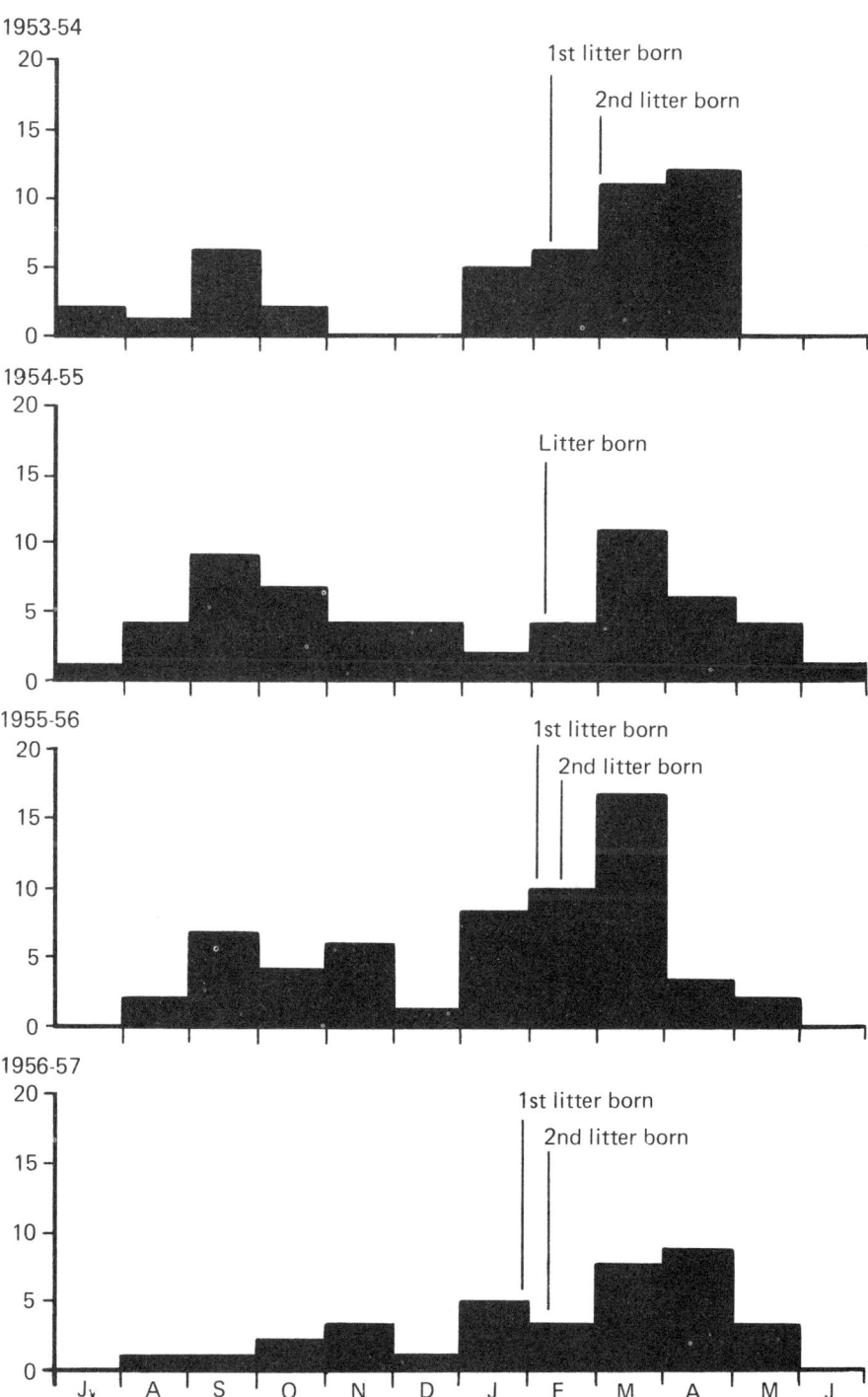

Figure 3.7 The relationship between bedding collecting and digging activity, based on their average frequencies for each month. Incorporating the same data as for Figures 3.5 and 3.6

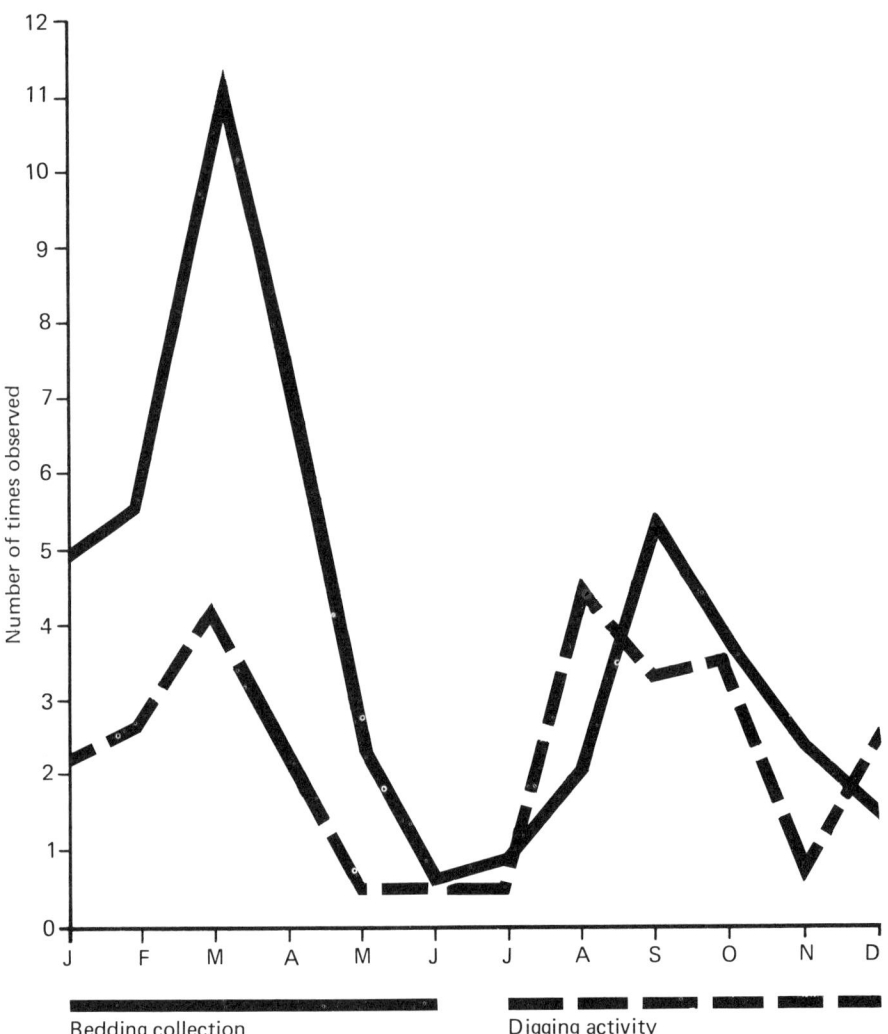

evidence that there will be cubs subsequently. What is more, the entrances into which it is taken appear to be the ones nearest the breeding chamber and also the ones the cubs will use during their early appearances above ground. I was able to use this correlation very usefully when planning to film a badger family with Professor Hewer. Lights had to be installed above the entrances well in advance of the time when the cubs would appear above ground in order to habituate the adults to the illumination. There were many possibilities, but by choosing the entrances into which bedding had been dragged in January we fortunately hit upon the ones the family subsequently used.

In northern districts of England, where litters tend to be later, this bout of bedding collecting is often several weeks later than in the mild south-west.

Neal and Avery found that much more digging and bedding collecting occurred during the February–April period when the cubs were confined below ground. Then the digging was largely concerned with the removal of old bedding which at this time is constantly fouled by the young cubs. However, some fresh digging did occur, particularly in February, but mainly at outlying holes. This was probably due to the fact that just around the time when the cubs are born the resident boars are usually banished from the breeding area and these clean out other parts of the sett complex.

Bedding collecting was a major activity in late February and March, and also in April in the years when there was a further litter born later. During this period it was collected on almost any suitable night—that is to say when it had been dry for at least two days.

In May little fresh digging took place, but in most years a lot of bluebell leaves were brought in when green.

The Significance of the Bedding

The major function of bedding appears to be concerned with preventing heat loss. This applies to adults when they are sleeping during the daytime, but to a far greater extent to the cubs during the first few weeks of their lives. When young the cubs snuggle right inside this mass of dry bedding. The heat generated in their bodies and that from the sow when she is nursing them is retained in the nest, as the bedding is an excellent insulator. There is no doubt that quite a high nest temperature is maintained in this way (p. 163). The dryness of the bedding and its presence in large quantities seem to be important factors for survival.

This is illustrated by the lengths to which some sows will go in order to provide dry bedding when conditions are difficult. For example, in one locality good use was made of the fibrous bark of Sequoia trees which was the only dry material available. This was clawed off and brought back a considerable distance.

Norah Burke (1964) considers that badgers sometimes draw sticks into the sett and sometimes pine cones to act as a sort of mattress between the damp earth and the bedding. She found in a chamber that she had excavated that the sticks below the bedding showed signs of having been chopped by the badgers into suitable lengths. Whether this is a deliberate action, or the pieces are brought in fortuitously along with the bedding remains to be proven, but the tooth marks on the ends certainly suggest that the badger uses the sticks for some purpose.

There is no doubt that sticks, especially dead elders, are brought back by badgers and it is not unusual to find long ones stuck in an entrance. Michael Clark once found a long stick projecting from the main entrance of a sett in a fir wood he was visiting during the day. It looked as if someone had just pushed it down in passing.

> I was about to pull it out, when I realised it was waving slightly in the air. It was then given a sharp pull and I saw it move further below ground. This continued, but it did not sink in much further because of the turn in the tunnel. Eventually I took hold of the waving end and gave a pull. There was immediate resistance, but after a few moments the tug-of-war with the badger was over and out came the stick. The far end showed numerous badger hairs and tooth marks!

Although most bedding is brought in dry there are exceptions. Enid Smith gave two instances of it being brought in after light snowfall, on nights when dew had been heavy, and several times after rainfall. Most badger watchers would agree that this happens occasionally in most districts, but is not typical and would not be used for young cubs. It was suggested by Enid Smith that some dampness is desirable to keep the nest sufficiently humid to prevent too much water loss from the badger's body. It is also possible that slightly damp bedding would ferment and thus increase the temperature of the nest in the same way as a heap of damp hay generates heat. Possible evidence for this idea comes from badger diggers who have commented upon the rotting condition of much of the bedding they have found at certain times of the year. The presence of green bedding in May might have a similar function, although Eric Ashby has suggested that this green material might be used by the cubs during the weaning period as supplementary food. This is a possibility, but I know of no evidence to support this hypothesis.

Badgers occasionally 'air their bedding'. This activity should not be confused with the odd bundle of dropped bedding near a sett entrance. It occurs most often on dry sunny mornings in winter. Great quantities of bedding—usually hay—are dragged out and fluffed up into loose piles which are scattered all around the entrances (Figure 3.8). After a few hours in the sun the badgers draw it back again. The only logical explanation of this is that the bedding is being dried.

DAY NESTS

In isolated districts badgers sometimes spend long periods above ground during the day. This happens mainly from midsummer until the autumn. Three kinds of day nests may be distinguished: those used for breeding; summer nests which are some distance from a main sett; and nests in close proximity to a sett.

Breeding nests above ground are unusual, but may occasionally be found in places where the water table is too high for digging or where, for some reason, a sow has been displaced from a more normal situation. These will be described in Chapter 10.

Summer nests often occur where badgers move to areas with more abundant food at that season. Special places are chosen for these, such as a hollowed out bramble bush, dense bracken or the centre of a wheatfield. Conspicuous badger paths often lead to these day nests and much bedding is brought to them, sometimes from long distances. Dung pits are often found nearby.

One of the most extraordinary summer nests I have seen was shown to me by Perarvid Skoog. It was in a hollow tree on one of the Baltic Islands and was regularly used when the whortleberries were ripe in that part of the island. Not only was it full of bedding, but this overflowed in a great heap outside (Plate 12). A second hollow tree in the same area was also used, but it did not contain such a spectacular amount of bedding.

Summer nests are not always associated with areas where there is an abundance of food. I found one such nest under rhododendrons on the Blackdown Hills and John Breeds found a similar one in Dorset 100 m from the nearest sett on the side of a patch of bog.

Badgers have been disturbed from day nests on several occasions.

Figure 3.8 Bedding being aired in winter

Blakeborough and Pease (1914) described how hounds came across a badger asleep in some deep bracken; the lair contained 'quite a cartload of dead bracken, well pressed down and so arranged as to shelter the sleeping animal'. They also mention another occasion when they were able to creep up to a day nest and to their delight see two adults lying there fast asleep.

It was not until recently that I was fortunate enough to see badgers using a day nest—an experience I wrote about more fully in *The Countryman* (Neal 1982). When making a routine visit to a favourite sett in Somerset on 19 June I found a heavy crop of hay had recently been cut and the badgers in the copse nearby had made the most of this superabundance of potential bedding. They had not only filled the chambers below ground, but outside one of the entrances was a huge mass of hay about a foot deep covering an area of about 2×1.4 m. Two hollowed-out regions showed where badgers had compressed the hay by lying on it.

A few days later when I arrived at 19.30 with the sun still high the nest was unoccupied, but shortly after, a boar emerged, went straight to the nest and thoroughly groomed himself in one of the hollows before leaving. A sow then emerged, used the nest for grooming for a few minutes and then retired below. Two yearlings then appeared but left the sett without taking any interest in the hay. Soon the sow re-emerged, this time with her well grown cub and at the same time the boar returned from his territorial duties. All three made a bee-line for the nest, the sow and cub used one hollow, the boar the other. After a long session of grooming, the sow took some hay from the edge of the pile and shuffled backwards with it down below. Later she went off with the cub leaving the boar in sole possession of the nest. It was some time before he eventually ambled off. Badgers had been in view for over two hours and most of the time their activities had been centred on that remarkable nest of hay.

A few evenings later I went again. This time the sow went off almost at once, but the boar made straight for the nest. No grooming this time. Instead, he pivoted round and round, put his head between his paws and prepared to go to sleep. With his muzzle deep in the hay he sneezed several times probably because dust particles tickled his nose and then went into a deep sleep. His respirations were slow—ten or eleven deep breaths per minute. I took several flash pictures, but he just went on sleeping. After more than an hour's sleep I moved the camera and tripod to within 2.5 m for better pictures, but still no response. Eventually I retreated to 5 m and woke him up by clicking my tongue. He shot bolt upright and stared straight at

Figure 3.9 Badger asleep in a day nest

me as if he couldn't believe his eyes. Even after another photograph he continued to stare; then without hurry or even a backward glance he slipped away into the bushes. I was sorry to have disturbed his dreams, but felt very privileged to have been present on such an occasion.

Shallow hollows are also seen near sett entrances and sometimes they may contain bedding. I have seen badgers use them when grooming soon after emergence and occasionally two animals will use the same hollow.

PATHS

In the vicinity of an occupied badger sett, there is usually a well marked system of paths. These paths, relatively few in number, lead from the used entrances to places of importance within the home range of the resident badgers, such as alternative setts, main feeding grounds, dung pits and drinking places.

These paths are often very distinct and may be quite denuded of vegetation due to the constant trampling. The main paths may be followed in woodland for 300–400 m without difficulty, and people often use them without realising their origins until the path passes under bushes or the lower branches of trees. One is tempted to believe that some of the ancient human trackways had their origins in animal tracks such as these. Where the undergrowth is thick, they take the form of tunnels just as they do when penetrating a hedgerow. Where a badger path leaves a wood badger hairs may often be found caught in the lowest strand of barbed wire. Occasionally, instead of finding the characteristic straight wiry guard hairs with normal colour pattern, you find much finer black ones. This happens when the badger passes over the wire rather than under it, so the ventral hairs become caught!

In woods, the paths may meander considerably according to the lie of the land and the position of trees and other barriers to a badger's progress. They usually represent the easiest route to travel. It is probable that badgers follow these tracks largely by scent, the snout being held just above ground level in order to recognise the scenting places marked by badgers along the

Figure 3.10 Leaving hair on barbed wire above a path

THE BADGER'S HOME AND ENVIRONS

Figure 3.11 A deciduous wood in the Quantock Hills, Somerset, containing four small setts belonging to the same social group. Not all setts were occupied continuously

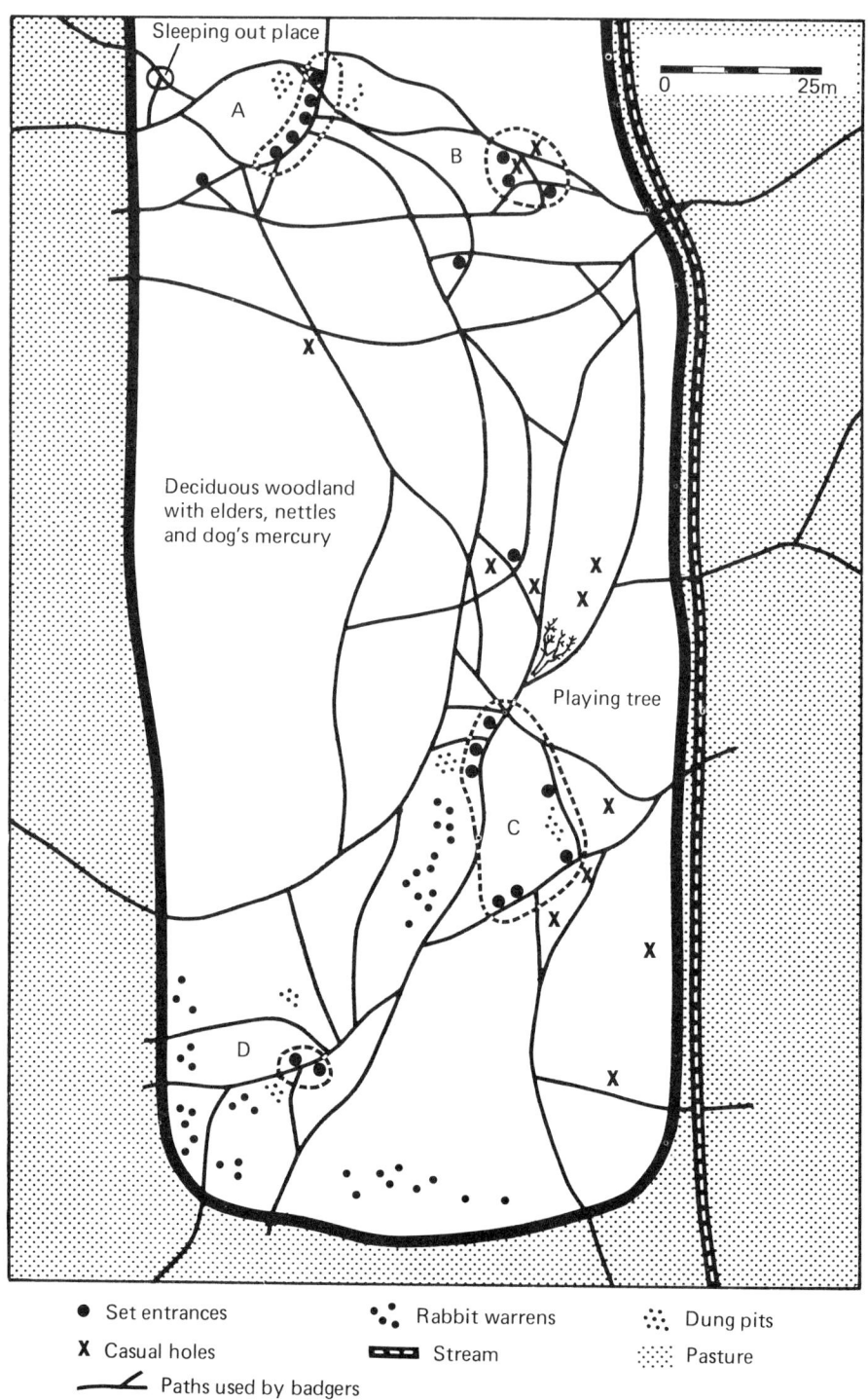

THE BADGER'S HOME AND ENVIRONS

path on previous occasions (p. 144). If you stand on a path and a badger comes towards you it may almost blunder into you before realising, by scent, that you are standing in the way. However, near to the sett, particular landmarks such as trees, sudden twists in the path and other characteristic features may help a badger to pin-point its exact position. Humphries (1958) suggests that a badger's knowledge of a path system is related to maze learning in other animals. This would seem very probable as a badger, if frightened near a sett, rushes for home along one of these paths rather than by a possibly shorter route 'cross country'. The learning of a path system would certainly have survival value by enabling the animal to find the safety of the sett quickly.

The path system near permanently occupied setts changes very little over the years. In a wood in the Quantocks which I have kept under observation for nearly 30 years the few changes in the paths were merely due to the falling into disuse of one of the four setts in the wood and to the excavation of new holes.

If a sett is near pasture the badger paths, especially in early spring before much grass has grown, can easily be seen going some 30–40 m into the field before becoming indistinct.

Long established paths which link setts together or lead to main feeding grounds often pass across man-made barriers such as lanes, main roads or railways. In the sunken lanes of Somerset and Devon you can often find their 'up-and-overs'. These are bare slides down the steep banks which badgers use when crossing the lane (Plate 13). If you find one on one side there is usually another one not quite opposite on the other. Similarly if

Figure 3.12 Badgers often clamber over a dead tree trunk if it straddles a main path

THE BADGER'S HOME AND ENVIRONS

there are dry stone walls, as in the Cotswolds or Cornwall, you can sometimes see that a few stones have been dislodged where badgers have climbed over. The moss is also usually worn away at such places (Figure 3.14).

Where the home range is very large, badger paths can be traced over very long distances. In Sweden, Skoog showed me a path which led from the sett area across a bare rocky region where little food was to be found to a place 1 km away where food was seasonally plentiful. Where the path crossed the rocky area, for several hundred metres you could trace it quite easily because the grey lichen, *Cladonia*, which covered the rock was completely

Figure 3.13 Badger path from wood into pasture; March

THE BADGER'S HOME AND ENVIRONS

Figure 3.14 When badgers clamber over stone walls, they regularly use the same place, wearing away the lichens in the process

worn away by badger traffic. Badgers in Cornwall are also known to use paths which traverse the sea shore as a short cut between feeding grounds.

When roads are enlarged or new ones constructed badgers are often killed when trying to cross. Many years ago I had an extraordinary illustration of this. A taximan called at my house in Somerset when I was having breakfast one February morning and said he had a dead badger in the taxi and would I like it? He told me how he was taking a passenger to catch an early morning train while it was still dark. Just outside Taunton he rounded a sharp bend and went straight into the badger and killed it. It was a sow. The incredible thing was, that some days later he turned up again with another badger which he had killed at the same place when taking the same man to catch the corresponding train! This time the badger was a boar. The coincidence seemed to be remarkable, but the explanation must have been that both animals had lived in the same sett and used to cross the road at the same place just before dawn on their way home. It was unfortunate for the badgers that the taximan was also keeping to time.

DUNG PITS

Some dung pits are usually found quite near a main sett, others occur at strategic places in other parts of the home range. Newcombe (1982) found that for 55 dung-pit sites near setts 56 per cent were within 10 m and 43.4 per cent within 5 m. These pits are usually dug where the soil is soft, often under a bush, a fallen tree trunk or an overhanging rock. The pits are funnel-shaped, about 150 mm deep and roughly the same diameter at the top. When a pit becomes full of dung, it is left uncovered and another dug nearby. You often find five or six in the same area. After some weeks of use, these pits may be abandoned and another site chosen where new ones are dug.

Occasionally, dung is deposited in a disused sett entrance and I have known it on the spoil heap outside an occupied one. But this was after a night when cattle had trampled all over the sett area, and possibly the badgers had been too frightened by the invasion to leave the sett for their normal dung pits.

Dung pits near the sett are used most often in the late winter and early spring and it seems probable that at this season they are mainly used by the sows and cubs. Cubs use these pits within a few days of their first appearance above ground, so this provides a useful indication of when this event takes place. As the cubs use different pits from the adults, their droppings can easily be recognised.

With some setts, even large ones, you can find no trace of dung pits nearby and some have this peculiarity year after year. This does not appear to be due to the presence or absence of cubs, so presumably these badgers must defecate below ground: this happens commonly during the winter. Some excavated setts have revealed large quantities of dung, usually deposited in side chambers. Perhaps in some setts the winter habit persists.

Dung pits may also be found a considerable distance away from a sett. Some of these have territorial significance and are used to demonstrate by scent signals the ownership of the territories. Some are situated beside main badger tracks, especially where they reach some obvious landmark such as a hedgerow, a lane or the edge of a wood. Others appear to mark the perimeter of the territory of a social group and may be found in large concentrations.

As many as 50 pits within a relatively small well-defined area were noted on Bredon Hill in Worcestershire (Pickvance 1960) where there was a high density of badgers. The importance of these communal dung pits can be gauged by the prominence of the path leading to them. Mrs Harfield, an archaeologist who investigated the Iron Age fort of Cadbury Castle in Somerset, told me how the badgers living there on the western slope of the hill had made a track at least 400 m long which crossed a cornfield on the top and ended at a communal dung pit area on the other side. The path was so well marked that it could easily be seen in aerial photographs. The significance of these territorial dung pits will be further discussed in Chapter 9.

Not all dung pits situated some way from a sett are primarily territorial in function. This is true of those found in special feeding areas. For example at Hodd Hill, in Dorset, when the yew trees drop their berries, the badgers gorge themselves on these delicacies for a few weeks and dung pits are made nearby. They are also made in cornfields at harvest time and wherever there is an abundance of a particular food. It is probable in such feeding areas that the dung serves as a claim to ownership of the harvest. Sometimes, it may lie on the surface of the ground—not in pits.

Paget and Middleton (1974a) have commented upon the fact that some pits are only used for urination and Eric Ashby made the same point. When watching a sett in the New Forest, Ashby described how a badger emerged, went about 20 m to some empty dung pits and urinated, after which, taking a step forward it wiped its feet with long brisk scrapes like a dog. It then walked another 20 m or so to some dung pits containing droppings which it

THE BADGER'S HOME AND ENVIRONS

used for defecation, again scraping its feet. These actions may also have territorial significance (p. 143).

SCRATCHING TREES

It is usual to find near one of the entrances of a main sett a tree which shows obvious signs of claw marks and mud up to a height of about 1 m. These scratching trees are usually elders but other rough-barked species such as oaks and elms are occasionally used.

When a badger uses such a tree, it gets up on its hind legs, reaches as high as it can with its front paws and then brings them down, scraping them against the bark as it does so. The bark may eventually become scratched away by this action, and if the soil is coloured, the mud clinging to the trunk is very conspicuous.

This action has been described as claw sharpening, but that is clearly incorrect as no such action will sharpen the claws; rather the contrary.

It is not at all certain why badgers do this. It has been suggested that it is to tone up the muscles after sleeping, as many mammals stretch themselves on waking; but although it can happen shortly after emergence, it may occur after returning to the sett just before dawn.

A more likely explanation is that after a night's foraging or a bout of digging, mud may collect between the toes and this action helps to remove it. It is certainly my impression that such trees are used more often when the ground is wet and muddy, but not exclusively so. It could be significant that

Figure 3.15 Scratching an elder tree by the sett. Note the shredded bark

scratching trees are seldom found near setts dug in very light sandy soil, as light sand would more easily become dislodged from between the claws without having resort to scraping.

Another possible explanation is that a scratching tree serves as a visual or scent signal which would have territorial significance. It is known that the wolverine, *Gulo gulo*, which is a near relative, marks trees near its den, but this appears to be due to repeated chewing and biting. Consequently, the trees become very conspicuous and may function as visual signals. Unlike the wolverine, badgers are mainly nocturnal so a scratching tree is unlikely to act as a visual signal for them, but it may well act as a scent marker. Many mammals have scent glands between their foot pads and several species of the cat family scratch at particular trees and combine this action with rubbing the body against it and sometimes urinating. I have not noticed badgers rubbing against such trees, but nevertheless, some with hard bark show the mud almost polished as if rubbing had occurred. It is also true that in places where suitable trees are not available near a sett, you sometimes see mud marks on boulders or similar structures nearby. You also find scratch marks on fallen tree trunks which indicate repeated rather than random use.

It is clear that more observations are needed before the full significance of these 'scratching' places is understood.

4 Distribution and choice of habitat

DISTRIBUTION

There is only one species of the Eurasian badger, *Meles meles*. Its distribution is very wide, stretching from Ireland in the west, right across Europe and Asia to Japan. It is found practically all over Europe where local conditions are still suitable, except for the northernmost parts of Scandinavia and Russia. It is still fairly common in Spain and Portugal and in the hillier parts of other countries bordering the Mediterranean and is present on such islands as Rhodes and Crete.

It is widely distributed in the southern parts of Norway and Sweden and is fairly common in Denmark. The population in Holland although small, has risen in the recent past due to conservation measures (Wijngaarden and Peppel 1964).

In France and Germany, it is widely distributed and fairly common in the more wooded and hilly districts; the same applies to Switzerland, Austria and Poland. But in all these countries it is not so common as previously.

It occurs fairly commonly in the Balkan countries and is widespread in the USSR. It also occurs in the Middle East including Jordan and Israel.

In Asia, it is not found much north of a line from where the rivers Tabor and Ob meet to the north of Amurland. Thus, in west Asia, badgers are to be found near the Arctic Circle, but in the east the northern limit is much further south.

Eurasian badgers are present on several of the Japanese islands and in most of China as far south as Hong Kong, with the exception of the higher parts of west China. The animal's southern limit is defined by the Himalayas, but it is not uncommon in Iran where it has been found at an altitude of up to 2,200 m.

Considering the tremendous extent of its range in Europe and Asia, it is not surprising that many different forms have been described which differ particularly in colour, size and minor dental and skeletal characteristics. At least 23 sub-species have been named (Ellerman and Morrison-Scott 1961), but a number of these are better looked upon as geographical races. Bobrinskii *et al* (1944) in their survey of the mammals of the USSR, group together ten of these races into four sub-species. These are: *M.m. meles*, the type species which is widely distributed over most of Europe; *M.m. canescens*, a much smaller badger from Transcaucasia; *M.m. leptorhynchus*, which has a wide distribution from south-east Russia and the Caucasus through much of Siberia; and *M.m. amurensis* from Manchuria.

Further Asiatic sub-species include *M.m. anakuma*, which is confined to Japan. This is a small, brownish badger with a chocolate-brown eye stripe, which in some specimens is reduced to a ring round the eye, giving the animal a panda-like appearance. It is largely a forest species which is found up to an altitude of 1,700 m. *M.m. leucurus* is another sub-species found largely in China and Tibet and *M.m. albogularis* is also found in Tibet.

Some island forms have also been given sub-specific rank. These include *M.m. rhodius* from Rhodes and *M.m. arcalis* from Crete. These appear to be relatively small species.

In western Europe, the only sub-species for which there is any justification are those from Spain and Portugal, *M.m. marianensis* and the Danish badgers, *M.m. danicus*. The latter, according to Dahl (1954), have significant differences from the type species in the teeth and skull.

CHOICE OF HABITAT

Badgers inhabit a very wide range of habitats. In Britain their setts are found in woods and copses, scrub and hedgerow, in quarries and sea cliffs, in moorland and mountainous country, in open fields and downlands, in green belts within city boundaries and on housing estates. They also occasionally make their setts in the embankments of roads, railways and canals, in long barrows and Iron Age forts, in mines and natural caves, in coal tips and rubbish dumps, and even under buildings and major roads. This extraordinary adaptability has undoubtedly contributed largely to their success as a species.

In 1963 the Mammal Society organised a National Badger Survey, one of the aims of which was to find out the important broad ecological factors which had a bearing on the distribution of badgers in Britain and those local factors which influenced the badgers' choice of site for a sett.

In the original survey, some 5,000 setts were documented by County Recorders and their helpers from many parts of Britain (Neal 1972a). Since that time the tally of setts analysed in detail has reached nearly 12,000. This later material has been incorporated in Table 4.1 which summarises the results.

With such a diversity of habitats it is often impossible to separate out the relative importance of the various ecological factors as they are obviously inter-related and differ in priority from one habitat to another. It should also be borne in mind that much more reliable conclusions can be drawn from data concerning main breeding setts rather than outliers, as many of the latter are used only seasonally.

With these points in mind some of the more important factors which influence choice of site will be discussed. Any percentages quoted refer to data from the Survey.

Geological Factors

The type of soil is of paramount importance in the choice of site. The ideal sett appears to be one which is easy to dig, is dry and therefore relatively warm, and is safe from predators and the danger of the roof collapsing. For these reasons a sandy soil is normally preferred to one of clay, although the latter is not despised in districts where there is little choice. However, very heavy clays are avoided.

Setts in sand are predictably much more extensive than those in other soils and enormous workings have been found in suitable greensand and red sandstone regions. The only disadvantage of sandy soils is that if they are very light there is a risk of the roof collapsing. However, these soils are often associated with woodland and badgers choose the regions where tree roots provide the needed strength. Exceptionally, badgers form setts in consolidated sandhills near the coast; here the old rhizomes of marram grass often provide support. This was so in a sett almost at sea-level on the coast of Argyll, Scotland and also in Coto Donana in southern Spain. They have also been reported in similar situations in Russia (1956).

The preference for sand over clay is well illustrated by the distribution of setts in the Weald of Sussex, an area where the badger population density is very high. E.D. Clements (1974) describes how in this area layers of sand and clay alternate to give the badgers a good choice. However, almost all the setts are to be found in the Ashdown sand and Tunbridge Wells sand, very few in the clay. He goes on to say that 'both these sandy areas are very variable in character, ranging from pure sand in places to a very clayey texture in others. Again, the badgers select the sandy spots.' Further details are given in Table 4.2.

Similarly, in the Greater London area, only two setts were found in the London clay compared with 84 in sand and gravel and 45 in chalk (Teagle 1969).

Wytham Woods, near Oxford, cover an area of hard coralline limestone at the higher levels and heavy clay on the lower slopes, with a belt of calcareous grit sand between. All the eight major setts in the woods are found in this sandy belt (Southern 1964) and on neighbouring hills all the setts also follow the line of the same sandy stratum (Hancox 1973).

Chalk is also popular with badgers. This gives excellent drainage and the setts are extremely well protected by the hard rock. However, some chalk is very hard and difficult to excavate and it is evident where a choice is available that they choose the softer strata, or where the chalk is broken up into more easily manipulated lumps, or is mixed with flints which can be dislodged. At Box Hill, Surrey, the large badger setts are dug in soft, sandy, chalk and you find many flints thrown out on the spoil heaps. In the Chiltern Hills in Buckinghamshire a study by M.R. Dunwell and A. Killingley (1969) of the distribution of setts in relation to the geology found only one sett on the extensive clay plateau compared with 117 in the chalk. Of the latter 94 were in upper chalk and 23 in middle chalk. They point out that the stratum of upper chalk often coincided with bands of woodland and that this probably explained the preference for upper chalk.

Further evidence of preferences can be obtained by studying an area where the badger population is increasing and badgers are overflowing to less suitable areas. Beatrice Gillam (1967), describing such an area in Wiltshire, comments that typically the setts are mainly in the wooded upper greensand slopes where digging is easy, but when this first choice habitat is fully occupied, the surplus badgers do not move down to the gault clay, but up to the chalk and finally to the open unploughed downland. However, these choices are not simply those of geology as other factors, such as adequate cover and seclusion, must also be important. These factors will be discussed later.

Table 4.1 Percentages of setts in various habitats in selected counties of Great Britain

England	Deciduous and mixed woods, copses	Coniferous woods	Hedge	Scrub	Open field
Avon	61.0	2.4	19.6	—	10.6
Bedfordshire	21.6	6.1	46.2	6.1	1.3
Berkshire	70.0	3.3	13.4	5.0	5.0
Buckinghamshire	66.4	0.3	17.6	1.1	8.9
Cambridgeshire	37.8	4.4	20.9	10.7	10.7
Cheshire	72.1	3.8	6.3	1.3	11.3
Cornwall	38.6	6.0	4.0	25.1	4.8
Cumbria	40.2	6.2	1.8	—	16.1
Derbyshire	38.6	12.2	7.3	11.0	8.1
Devonshire	61.8	10.3	9.6	8.9	4.1
Dorset	39.2	5.0	14.8	15.4	20.9
Durham	59.0	21.8	2.6	9.0	—
Essex	62.9	1.5	20.7	3.5	1.3
Gloucestershire	75.0	6.3	9.4	3.1	3.1
Hampshire & I.o.W.	60.8	14.9	4.5	8.0	2.2
Hereford & Worcester	54.5	9.5	8.0	1.5	15.0
Hertfordshire	55.3	5.2	14.9	6.0	3.7
Humberside	48.0	14.0	6.0	4.0	16.0
Kent	64.2	4.2	8.5	7.8	8.1
Leicestershire	46.9	1.8	12.6	5.4	14.4
Greater London	74.2	—	12.2	7.2	0.7
Norfolk	30.9	41.7	4.8	1.2	—
Northamptonshire	39.8	1.2	15.7	9.6	24.1
Northumberland	57.9	20.4	4.0	6.7	4.0
Nottinghamshire	50.0	9.0	16.0	2.0	4.0
Oxfordshire	58.1	2.3	15.1	2.3	4.7
Shropshire	62.0	24.5	3.9	1.1	1.7
Somerset	46.1	5.8	21.5	5.4	10.9
Suffolk	22.3	11.8	30.6	10.4	2.1
Surrey	81.1	4.8	6.3	3.7	0.6
East Sussex (Downs)	33.8	—	5.2	44.0	14.0
East Sussex (Weald)	66.4	4.1	17.4	6.6	1.7
East Sussex (Total)	56.5	2.8	13.7	18.0	5.5
West Sussex	74.4	2.5	8.3	9.4	3.6
Warwickshire	56.5	—	2.8	9.0	17.9
Wiltshire	31.3	4.8	9.7	6.0	40.9
North Yorkshire	49.2	19.7	2.3	8.5	8.6
South Yorkshire	46.0	12.0	16.6	1.3	6.7
England Total Per Cent	53.9	7.2	11.9	8.7	9.0
Wales					
Clwyd	59.4	13.0	7.7	5.7	3.8
Dyfed	28.9	3.8	21.2	17.3	1.9
Wales Total Per Cent	56.7	12.2	8.9	6.7	3.7
Scotland					
Borders	31.6	28.8	10.8	1.1	7.3
Dumfries & Galloway	18.9	19.9	10.0	1.0	8.4
Grampian	18.9	58.2	1.3	6.8	2.7
Highlands	34.2	16.6	1.6	2.4	3.2
Lothians	58.1	11.5	8.1	3.0	7.7
Strathclyde	57.4	9.6	8.8	2.9	4.4
Tayside	26.2	34.4	—	8.2	4.9
Scotland Total Per Cent	37.4	22.0	7.5	2.7	6.3
Grand Total Per Cent	52.5	8.8	11.4	8.0	8.5

DISTRIBUTION AND CHOICE OF HABITAT

Quarry	Sea cliff	Moor	Built-up area	Any other	Setts involved
2.4	0.8	0.8	1.6	0.8	123
4.6	—	—	—	3.1	65
—	—	—	—	3.3	60
4.1	—	—	0.3	1.3	370
12.9	—	0.4	—	2.2	225
1.3	—	1.3	1.3	1.3	79
1.2	12.3	6.4	—	1.6	251
6.2	—	28.6	—	0.9	112
4.9	—	4.5	—	13.4	246
2.9	0.2	2.2	—	—	417
2.1	1.1	—	t.4	1.1	526
3.8	—	—	—	3.8	78
4.3	—	—	3.5	2.3	395
—	—	—	—	3.1	64
2.2	0.5	5.3	—	1.6	375
6.5	—	—	—	5.0	200
14.9	—	—	—	—	134
8.0	—	—	—	4.0	50
4.0	0.1	—	0.2	2.9	828
7.2	—	—	—	11.7	111
1.4	—	—	2.9	1.4	140
11.9	—	9.5	—	—	84
4.8	—	—	2.4	2.4	83
2.7	0.4	2.2	0.4	1.3	225
1.0	—	2.0	1.0	15.0	100
14.0	—	—	—	3.5	86
1.7	—	2.8	0.6	1.7	179
6.1	—	1.7	—	2.5	762
18.1	0.4	0.4	—	3.9	287
2.2	—	0.2	0.9	0.2	539
0.7	—	—	1.6	0.7	443
2.7	0.2	—	0.4	0.5	1,004
2.1	0.1	—	0.8	0.5	(1,447)
0.7	—	—	0.4	0.7	277
6.9	—	—	—	6.9	1458
1.7	—	—	0.2	5.4	518
3.6	—	5.7	—	2.4	614
6.7	—	8.7	—	2.0	150
4.2	0.4	1.7	0.5	2.5	10,345
3.1	—	1.6	—	5.7	546
—	17.3	3.8	—	5.8	52
2.8	1.5	1.8	—	5.7	598
0.4	—	14.6	—	5.4	260
1.0	26.7	7.8	—	6.3	191
—	8.1	2.7	—	1.3	74
—	—	40.4	—	1.6	126
5.2	—	0.9	2.1	3.4	234
—	2.9	11.8	—	2.2	136
—	—	19.7	—	6.6	61
1.4	5.6	12.5	0.5	4.1	1,082
3.9	1.0	2.7	0.4	2.8	12,025

Table 4.2 Number of setts in various geological strata in Sussex

Soil type	Number of setts	Per cent
Clay with flints	4	
Upper/middle chalk	458	27
Lower chalk	85	5
Upper greensand	1	
Gault clay	0	
Lower greensand	4	
Folkestone beds	14	1
Sandgate beds	10	
Hythe beds	43	2.5
Atherfield clay	0	
Weald clay	13	
Sand/sandstone in Weald clay	11	
Tunbridge Wells sand	276	16
Grinstead clay	0	
Ardingly sandstone	69	4
Wadhurst clay	7	
Sand in Wadhurst clay	54	3
Ashdown sand	583	34
Purbeck beds	2	
Head	46	2.5
Alluvium	21	1
Miscellaneous	18	1
Total	1,719	

Source: From Clements (1974, updated).

Limestone regions also have high badger populations. This is particularly true of the Cotswolds, the Mendips, the Pennines and parts of Yorkshire. Limestone gives good drainage and protection, but in some places the rock is too hard to excavate and the badgers then make use of local regions of more friable rock, natural fissures and cave systems. Speliologists have often come across huge piles of bedding on dry ledges in such caves or in small side caverns.

In many limestone regions there are strata of soft material below the solid limestone and badgers exploit the juncture between them as the hard rock forms a strong impermeable roof to the sett keeping it dry and preventing the roof from falling in, while the actual digging is easily done in the sand. Thus in the Mendips the badgers excavate the sandy stratum immediately below the hard carboniferous limestone, in Yorkshire they use the Permian sand just under the magnesium limestone and in the Cotswolds, the lias sands with the hard oolitic limestone as a roof.

D.C. Findlay (1973) made an excellent study of the distribution of setts on the Cotswold escarpment which illustrates this preference very well. Six strata were identified (Figure 4.1). Setts were most abundant in the well drained lias sands (3), but particularly so near the junctures with the oolitic limestone above. When the population increased and this stratum became

DISTRIBUTION AND CHOICE OF HABITAT

Figure 4.1 Distribution of setts in relation to soils and landscape (Cotswold Scarp). 1. Great oolite limestone; 2. Fuller's earth, clay; 3. inferior oolite limestone; 4. Cotswold sand; 5. Dyrham silts; 6. lower lias clay. After D.C. Findlay (1973)

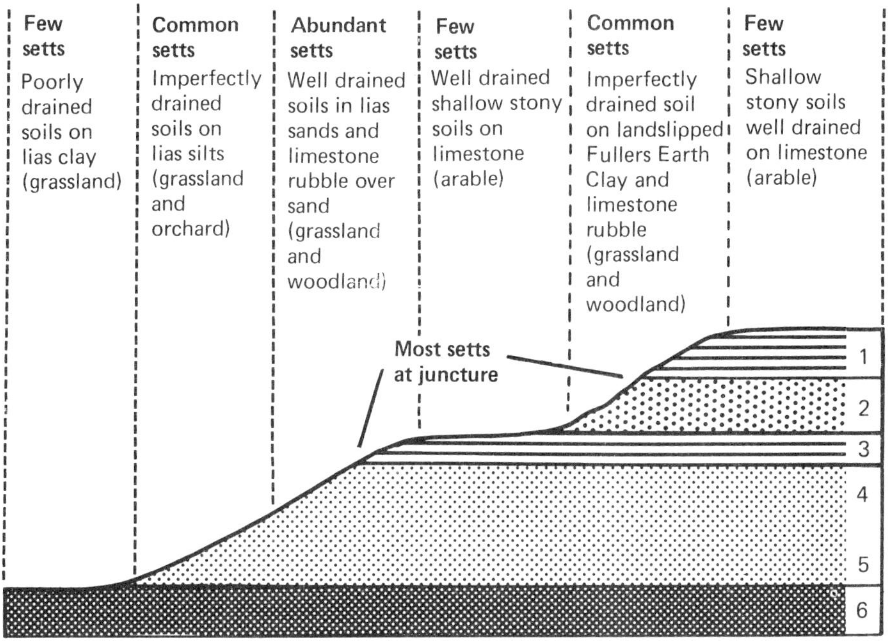

Few setts	Common setts	Abundant setts	Few setts	Common setts	Few setts
Poorly drained soils on lias clay (grassland)	Imperfectly drained soils on lias silts (grassland and orchard)	Well drained soils in lias sands and limestone rubble over sand (grassland and woodland)	Well drained shallow stony soils on limestone (arable)	Imperfectly drained soil on landslipped Fullers Earth Clay and limestone rubble (grassland and woodland)	Shallow stony soils well drained on limestone (arable)

saturated, it seems probable that they exploited clays higher up (2). Here the disadvantages of this sticky soil were to some extent offset by choosing in many instances those places where limestone scree overlaid it. The other option was to go downhill to the lias silts (5) where, although the soil was less well drained, there was more food available, especially earthworms.

The exploitation of soft strata with hard impermeable overlays also applies to other formations beside limestone. In the Blackdown Hills, Somerset, there is a long line of setts on the northern escarpment which are dug into the soft greensand leaving the hard, flint-like churt as a roof. Also in the Brendon Hills, Somerset, the new red sandstone is extremely popular, especially when the setts are roofed by bunter conglomerate. In Surrey, Clements reports that there are many setts at the juncture between the sandy Hythe beds and the hard sandstone rock of the Bargate beds. In Cumbria, many of the setts are excavated in softer soil below huge boulders, while others utilise the shelter of the stone walls high up on the Fells, the entrances occurring on both sides of the walls. In Yorkshire some are made under slabs of gritstone. Man has also provided similar opportunities for badgers to exploit. A sett in South Yorkshire, near Sheffield, was dug under a much-used main road, and in spite of the vibration caused by traffic above and the lack of cover on emergence, the badgers appeared to thrive. Badgers have also dug under sheds in gardens, under old lime kilns and even beneath the floor of a Roman villa.

This habit of choosing the juncture between two strata with contrasting properties may well account for the number of setts found in disused quarries, where the juncture is easily discovered. In many instances the hard

Figure 4.2 Sett dug in soft stratum under hard, impervious slates

rock is quarried out to the full depth of the stratum, so the juncture is at ground level. Badgers often dig in at this point. However, in some quarries in the Brendons in Somerset, both the red sandstone and the conglomerate layers have been used commercially—and in some of these the badgers have formed perilous paths down the quarry face to reach a ledge and have dug into the softer sand below the conglomerate. Piles of excavated sand at the base of the quarry face bear witness to their labours. Abandoned quarries may also be popular because of the vegetation cover that soon grows up and the relative freedom from interference.

Ease of digging probably accounts more than any other factor for badger setts being dug into places where man has disturbed the soil, possibly many years before, and left it less consolidated than other places nearby. If these places rise above the level of the surrounding country badgers find them even more attractive. It is probably for these reasons that setts are dug into road, railway and canal embankments, rubbish tips and prehistoric earthworks.

This habit has on several occasions been helpful to archaeologists. Henry Cleere wrote to me in 1966 about an eleventh-century Roman iron-smelting site near Ticehurst, in Sussex, which he had been excavating. On the site was a very large slag and rubbish bank covering 100 m of a small valley. He found a sett in this bank which the badgers had found easy to excavate in contrast to the natural sticky clay elsewhere. On turning over the loose soil of the spoil heap, he found that the badgers had unearthed some excellent shards of pottery including a magnificent piece of a Samian red-glass mortarium. Further excavation of the spoil heap showed that they had also pulled out some valuable pottery 'wasters', confirming Henry Cleere's theory that pottery had been manufactured on the site.

Long barrows provide the same sort of conditions and some of these in

Wiltshire are riddled with badger workings. The same applies to many ancient earthworks such as Cadbury Hill, Somerset and Wallbury Dells, Hertfordshire. Setts made in coal deposits occur in Northumberland and Yorkshire: Paget and Middleton (1974b) in their admirable account of the setts in Yorkshire, describe how in one sett in the parish of Cawthorne, the spoil heaps outside the entrances contained large lumps of coal mixed in with small coal and shale—the sett, in fact, being in an old pre-mechanisation mine working known as a day-hole.

Setts are also found in some of the Cornish mines. In one of these, badgers used to emerge from a low level mine entrance on to the beach below.

Soils which are periodically flooded or are liable to become waterlogged are avoided, but if the land rises sufficiently to give the necessary dryness for the sett itself the region may be exploited, as they are often useful feeding grounds. In the fenland district around Goole, Humberside, Richard Paget told me that the only setts in this very flat region were in the banks of dykes and railways. One of these setts extended along the bank of a dyke for about 100 m, the whole area being covered by elders. In the Somerset peat moors there are no setts on the flat areas where the water table is usually high, but if the land rises 3 m or more, setts may be found. An unusual site in Sussex was described to me by E.D. Clements. This was on the Kent marshes between Winchelsea and Rye. It was on a dry bank which had a dyke on one side and a river on the other and there was no cover whatever. In addition, there was a main road just beyond the river.

Occasionally such setts may be flooded out by freak weather conditions. On Canvey Island, Essex there was a thriving badger sett on the bank of one of the ditches. When the sea wall was breached and water flooded this area on 31 January 1953, it was thought to have drowned all the badgers. However, at the end of March, a farmer, F.J. Leach, noticed that some mounds of fresh sand had been excavated from this sett and that there was a trail of hay from a nearby haystack leading to the entrances. In fact, cubs were reared there successfully. How the badgers survived the flooding is not certain. They may have climbed up amongst the woodpiles of a neighbouring woodyard (Cowlin 1972) or found shelter in a haystack, or on slightly higher land nearby, but as the latter was very exposed, this is less likely. It is just conceivable that the sow survived in a chamber below ground, which because of its higher position relative to that of the tunnel contained a substantial amount of trapped air.

Cover

Some kind of cover near a sett is an important factor in choice of site. Cover allows the badgers to emerge inconspicuously, and it allows young cubs to play near the sett entrances without being visible to people or potential predators.

This need for cover is probably the major reason why 52 per cent of the 12,000 setts investigated in the 1963 Survey were found to be in deciduous woods and copses. If setts in hedgerow and scrub, which also provide good cover, are added to the number, it rises to 71 per cent. These figures are for the whole country, but for counties such as Surrey or Sussex it is about 90 per cent.

Where woods and copses were scarce, hedgerows and scrub were the most commonly used alternatives. Setts in hedgerows were much more common in Somerset and Devon, where traditionally they are set on banks, and if there were disused farm tracks bounded by banks with overgrown hedges, these were almost always appropriated.

Coniferous woodland contained comparatively few setts (8.7 per cent) and of these, the majority had been there when the original deciduous wood had been felled and replaced. It would appear that lack of ground cover and the scarcity of suitable food and bedding in the vicinity are the cause of their unpopularity. It is probably for these reasons that the setts which do occur in coniferous woodland are usually near the edge.

In the Survey, 9 per cent of setts were in open situations, but this rather surprisingly high figure should be treated with reservation, because the amount of cover in an apparently open situation varies greatly during the course of the year. In winter and early spring, a sett may appear to be without cover, but in summer it could be hidden in bracken. Other setts in open situations soon acquire cover as nettles, docks and thistles quickly colonise the excavated earth nearby.

An interesting aspect of this phenomenon is the seasonal movement away from setts for the duration of the exposed period. Thus, badgers in some coastal regions in Dyfed in Wales may use setts on the cliff slopes, where bracken and bluebells give cover in summer and autumn, but in winter and early spring use copses further inland.

This seasonal movement, however, is not typical of most bracken-covered regions, as times of emergence make this unnecessary. From September to the end of April, they come out after dark, so ground cover is less necessary, but from May to August, when the badgers often come out before it is dark, the bracken gives them cover. Moorland setts also come into this category, because although open, they often have good heather or long grass cover.

Setts in truly open situations are found more often at higher altitudes where disturbance is less and cover not so important. At lower altitudes open situations are often arable or pasture, so in the former they are subject to continual interference through farming practice and in the latter by the attentions of cattle. However, badgers sometimes do make setts in surprisingly open situations, even in flat fields. Such a sett, described to me by G.P. Knowles, was found in an open field which had recently been ploughed. It was invisible from the edge of the field as the excavated earth had been levelled out by the farm implements. Six holes were in continuous use. The farmer had been trying to get rid of the sett for 20 years, but had then given up as their tunnels led far down under the remains of an ancient Roman villa. In this instance, security seemed to be more important than lack of cover.

Some setts in open situations have their origins in hedgerows. When the hedgerows are removed to increase the size of the fields the setts may remain if secure and well established. Some hedgerow setts also have entrances some distance out into a field. These holes are seldom used for emergence, but they do offer an alternative route for re-entry.

Other open setts are to be found on downland. The majority of these are situated where some cover exists, but a few are in very exposed situations. There is one sett on a Derbyshire hilltop situated within the stones of a megalithic circle.

The type of cover found in the vicinity of setts varies enormously, according to the site. However, some species, notably elders (*Sambucus nigra*) and nettles (*Urtica dioica*) were found far more than any others. In some counties, over 70 per cent of the setts including nearly all the older established ones, had these two species nearby. This did not constitute a choice by the badgers, but represented the effect of their being there. There are several reasons for this. First, the dung and urine of the badgers make the nitrogen and phosphate content of the soil high, conditions very favourable to the growth of elders and nettles. Second, these plants can also thrive in soil which is subject to periodic disturbance, such as that produced by the constant digging; this is particularly true of nettles with their underground rhizomes. Third, and probably of most importance, badgers eat elderberries, the seeds pass unchanged through the gut and germinate in the old dung pits very near the sett. Finally, there is food selection by the rabbits which often live nearby, the elder seedlings being left because they are bitter and the nettles because of their stings.

This association of badgers with elders provides a useful means of finding setts quickly. When the elders are in bloom, their presence near the edge of a wood may easily be seen even from a distance. These regions should be the first to be searched.

A comparable association occurs with blackberries (*Rubus* sp.) in Britain, Holland and Germany as these are also a favourite food of the badger and their seeds too are indigestible. In Sweden, their place is taken by wild raspberries, which are a major food item for badgers in that country and their seeds germinate in a similar manner near the setts. Less commonly, rowan (*Pyrus aucuparia*) and occasionally gooseberry (*Ribes grossulariae*) occur too. The hemp nettle (*Galeopsis tetrahit*) also occurs fairly commonly near badger setts on the Continent and occasionally in Britain. The reasons are probably very similar to those for the nettle, the rank taste making up for the lack of stinging powers.

Slope

The great majority of setts are dug on sloping land (88 per cent). In some counties, such as Sussex, the figure is over 95 per cent. There is no doubt that this is advantageous to the badgers in a number of ways.

It is better to dig into a slope as this facilitates the removal of the excavated soil, which spills down the slope. It is also easier for a badger to find a particularly favourable stratum to dig into if it is on a slope, as it is more likely to be exposed. Sloping land is also usually well drained, so the sett is more likely to be dry and warm, and when digging into a slope, a depth below ground which is well below ground-frost level is rapidly attained.

One advantage resulting is that the mound which forms outside the entrance on a slope builds up to form a platform, where cubs and adults can remain soon after emergence. With continual excavations, the height of this platform increases above the level of the entrance, so the latter becomes the lower point of a kind of cone. This formation catches wind eddies from any direction, and a badger can detect danger without exposing itself fully.

Very steep slopes are often chosen where these are available. Chalk downland setts in Sussex are usually found near the top of very steep slopes (Clements 1974) and the sides of ravines are popular sites in mountainous country, such as in North Wales especially when these are wooded. It is probable that such setts provide greater protection from both weather and interference. Badgers greatly dislike cattle stamping around their setts and those in steep slopes tend to avoid this disturbance.

The direction of the slope appears to have little significance in most inland parts of Britain; but in exposed situations a preference is discernible. In Cornwall, 77 per cent of the setts faced away from the wet and prevalent winds (Bere 1970) and in Sussex (Clements 1974), within about 15 km of the sea, there is a tendency for them to face away from the prevailing wind which is strongest near the coast. In northern parts of Russia, Novikov (1956) states that setts are mainly on south-facing slopes of ravines where snow melts early in spring and there is no danger of flooding. In the south, however, such as in the Crimea and Caucasus, they prefer the less sunny northern slopes. In Britain there is a tendency in northern areas to choose the warmer southern slopes.

Altitude

In Britain, most setts are found at altitudes between 100–200 m. Several factors appear to be operating in determining this preference. Land below 100 m tends to be heavily cultivated or densely populated by man. Disturbance is less, at much higher altitudes, but here the food supply becomes too restricted, so medium heights are probably the best compromise.

Setts may occasionally be found almost at sea level if the site is not subjected to flooding and is secluded. One sett in north Somerset is in a shingle ridge not far beyond high tide mark and another, in Argyll, is in a sand dune a few metres above a sandy beach. Setts are also found at high altitudes. In the Yorkshire Dales many moorland setts are above 400 m, the highest reported by Paget and Middleton (1974b) at Buckden, being at 538 m. They are also found high up in Cumbria and the Highlands of Scotland and in south-west England near the highest parts of Dartmoor and Exmoor.

Scarcity of food seems to be the main factor which prevents badgers going very high, although in some mountainous regions the badgers are known to make nightly treks to lower levels where food is easier to acquire. In parts of Russia, Novikov (1956) mentions a seasonal migration downwards which may be related to available foraging as well as to avoidance of harsh conditions.

In more southern latitudes badgers are found much higher, setts having been reported from the USSR up to 4,000 m (Bobrinskii *et al* 1944).

Food Supply

The presence of a plentiful and varied food supply at all seasons of the year is undoubtedly one of the most important biotic factors influencing choice of site.

The diet of the badger will be discussed in detail in Chapter 8, but it must be explained here that although badgers are omnivorous and the range of food they eat is very wide, there are certain key items which are of vital importance. The most significant of these in Britain is the earthworm, *Lumbricus terrestris*, found abundantly in pasture. However, this major source of food needs to be supplemented by a wide range of other items, and these are more easily obtained from a varied, rather than a uniform countryside.

It is therefore not surprising to find that so many setts are found in places where woodland, grassland and arable are found nearby and in larger areas of deciduous woodland they are usually situated near the perimeter. In Shropshire (Russell 1967), 44 out of 49 of the setts recorded in woodland were within 25 m of the edge. This is generally true throughout Britain, although setts deep in large woods do occur. However, some of these border wide rides and most are found in regions where badger density is high, a factor which leads to the exploitation of less desirable sites.

Setts in copses, scrub and hedgerows bordering fields combine similar advantages of adequate cover and a varied food supply and these are all popular choices of site. It is therefore possible to speculate that the widespread availability of such habitats due to agricultural practice may well have contributed to the wide distribution of the badger in Britain.

Proximity of Water

In Britain, there are not many places where there is no source of water within about 1 km. Most setts are therefore within easy reach of a supply, so it is difficult to gauge its importance to the badger. Some setts are certainly situated very near to water, but in most cases this choice can usually be explained better by other factors. In drier country, such as upland calcareous pastures, badgers are known to visit cattle troughs and drink from the rain water which collects in the crotch between tree trunks. However, badgers are able to obtain much of the water they need through the food they eat, especially from earthworms and green vegetation.

Nearness to a Source of Suitable Bedding

This may be an important factor in the siting of some setts, although in most cases suitable material for bedding is easily available. For sleeping quarters, many kinds of bedding will do, but it seems probable that for successful rearing of cubs only certain kinds, especially hay and bracken, will provide the necessary insulation from loss of heat in colder climates. Thus, for a breeding sett, a near source of hay or bracken is an advantage. This is probably another reason why setts in woods, especially coniferous ones, are near an edge. Conifer needles make very poor bedding. Badgers living in setts deep in deciduous woods are better off, as the rides are often grassy and various herb layer plants make possible alternatives.

Seclusion

Setts are usually found in secluded places if the badgers have any choice, but nearness to habitation does not deter them from using an otherwise desirable site: it is human interference that disturbs them (Gillam 1967).

That badgers become tolerant of man's encroachment on their domain is obvious from the fact that 164 setts have been recorded within 30 km of the centre of London (Teagle 1969). Under semi-urban conditions they are almost a commensal of man. This aspect will be discussed further in Chapter 12. Meanwhile it is enough to quote from Clements (1974) concerning the badgers in Sussex towns.

> Hastings, especially, is well populated by badgers with one sett, which has trebled its size in the last ten years, in the centre of the town only 150 m from the sea. Little wonder that a badger was seen walking along the promenade at night. Another in Eastbourne dug the 19th hole of the promenade's putting green. There is also a 50-hole sett within the built-up area of Brighton.

If, however, a badger sett is repeatedly interfered with by man, the badgers will leave if they have an alternative sett to go to.

Analysis of the reasons for badgers leaving their setts in country districts illustrates this desire for seclusion and freedom from human interference. The first of these is tree felling. If woodland is clear felled, the resulting disturbance and lack of cover soon cause the badgers to abandon their setts and live elsewhere. This happened in Gloucestershire some years after I had done my original research on badgers in Conigre Wood. Although the badgers periodically returned to prospect, they did not return permanently until sufficient cover had once more built up.

Secondly, earth stopping by the hunt to prevent foxes sheltering in setts, if done too often, may cause abandonment. This applies more to outliers than to the well-established setts where interference is more likely to make the badgers wary rather than to cause them to leave.

Thirdly, repeated disturbance by cattle stamping round a sett may lead to temporary abandonment.

DISTRIBUTION IN BRITAIN

From the foregoing discussion on factors which influence a badger's choice of habitat for making a sett, it is clear that the distribution of badgers in Britain will to a large extent be determined by how well a particular area fulfils these ecological requirements. The main factors appear to be:

(1) A soil which is well drained, easy to dig and firm enough to prevent the roof from collapsing.
(2) The presence of an adequate food supply at all seasons of the year.
(3) Sufficient cover to allow the animals to emerge and leave without being conspicuous.
(4) A region in the immediate vicinity of the sett which is relatively free of disturbance from man or his animals.

The type of country which fulfils all these requirements is one which is hilly, has a sandy soil and contains a high proportion of deciduous woodland interspersed with fertile grassland containing a high density of earthworms. Many counties have some areas where this type of terrain occurs, but they are most frequent in the southern half of England, in the counties bordering England and Wales, in parts of Yorkshire, northern and north-western counties. These are the areas where badgers are most common.

However, there are few parts of Britain where badgers cannot exist at all apart from large conurbations, regions of high altitude and extensive lowlands which are liable to flooding. Being remarkably adaptable to a great variety of habitats their distribution is therefore very wide. While they are almost certainly present in every county of England, Wales and Scotland, their density is very variable and this aspect will be discussed more fully in Chapter 11. The reader is also referred to the map (Figure 11.1) in that chapter.

5 Sett tenants and associations with other animals

It is not surprising that with so much spacious accommodation provided by badgers, many other species make use of it. These include a great variety of invertebrates. Some, such as bumble bees (*Bombus* sp.) and wasps (*Vespa* sp.) attempt to make their nests near the openings of the tunnels. These often get destroyed by the badgers, but in unoccupied setts they may persist. Other invertebrates are associated with the bedding lining the sleeping and breeding chambers. These include badger parasites (p. 189), and also a large number of kinds of beetles, flies and mites. A list of these has been compiled by Hancox (1973).

Many kinds of mammals use badger setts in one way or another. The majority of these are casual visitors which make use of the tunnel system on a temporary basis, usually when the sett is unoccupied by badgers. This applies to pine martens, polecats, feral cats, wild cats, and several other small carnivores. Their occasional presence in a sett is of very little significance as far as the badger is concerned, as there is no interaction.

Other mammals may tunnel into the spoil heaps or make side burrows from the main sett. These include woodmice, bank voles and brown rats. It is somewhat incongruous when expecting a sizeable badger to emerge from a large entrance to see a tiny woodmouse appear on the threshold before darting off into the undergrowth! Brown rats use badger setts quite frequently in some districts, especially if the latter are near rubbish dumps.

In eastern Europe, wolves and racoon dogs occasionally use unoccupied setts as refuges (Novikov 1956) but not for breeding, as far as I am aware. However, the species which are commonly associated with badger setts, and which could be described as true commensals, are the rabbit and fox, over most of Europe, and the porcupine in Italy (Tinelli and Tinelli 1980).

Where rabbits are common, they are regular associates with badgers. Many setts have originated as enlarged rabbit burrows and in some, the rabbits have continued to use parts of the burrow system. Typically, in small setts rabbits take over as soon as badgers move out and you can be certain that if you find rabbit droppings near the entrance of a badger sett, no badgers will emerge from the hole. However, in large setts rabbits can live quite happily along with the badgers, but they use separate entrances and presumably live in tunnel systems of smaller diameter. Wijngaarden and Peppel (1964) mention a large sett in the Netherlands which contained a number of badgers from which no less than 50 rabbits were ferreted.

Table 5.1 Mammalian tenants of setts

English name	Scientific name	Status	Authority
Woodmouse	Apodemus sylvaticus	Casual	Hainard, Hancox, Neal
Bank vole	Clethryonomys glareolus	Casual	Hainard, Humphries
Brown rat	Rattus norvegicus	Casual	Barker, Hancox, Neal, Soper
Rabbit	Oryctolagus cuniculus	Commensal	Numerous
Fox	Vulpes vulpes	Commensal	Numerous
Wolf	Canis lupus	Casual in disused setts	Likhachev, Novikov
Porcupine	Hystrix cristata	Commensal	Tinelli and Tinelli
Racoon dog	Nyctereutes procyonoides	Casual in disused setts	Novikov, Popescu and Sin
Pine marten	Martes martes	Casual	Hancox, Tregarthen, Wijngaarden and Peppel
Stone marten	Martes foina	Casual	Jensen, P.V.
Polecat	Mustela putorius	Casual	Jensen, P.V.
Polcecat-ferret	Mustela furo	Casual	Paget and Middleton
Weasel	Mustela nivalis	Casual	Wood
Feral cat	Felis catus	Casual	Lancum, Wijngaarden and Peppel
Wild cat	Felis sylvestris	Casual	Crossland, Ognev

Up to the end of May, rabbits usually emerge and move off before the badgers appear, but in summer it is not unusual to see both above ground at the same time within a short distance of each other. Little notice is taken of the other, although the rabbits will bound off if a badger comes too near.

Badgers obtain young rabbits to eat by digging out their nests (p. 124), but I have never come across evidence of their doing so in the immediate environs of the sett in which they are living.

Foxes occasionally dig their own earths, often by enlarging rabbit burrows, but they are lazy diggers and much prefer to use badger setts if available. Typically, they make use of unoccupied setts where they are undisturbed, but in large setts they often share the spacious accommodation, one part being used by foxes and another by badgers.

In the Netherlands, Switzerland, Germany and Denmark, it is the rule for badgers and foxes to live in the same sett, the main exceptions being small, newly-formed setts which are occupied by badgers only. Wijngaarden and Peppel (1964) suggest that in the Netherlands this is due to the relative scarcity of suitable sett sites. There are many instances of fox litters in badger setts, and Likhachev (1956), over a 15-year period of research in Russia, found that in his study area 63 per cent of the fox litters were born in badger setts.

Paget and Middleton (1974b) describe how, in Yorkshire, badgers and foxes often live together in man-made sites, such as old coal mine workings. They state that gamekeepers regularly cause foxes to bolt from these setts in some numbers.

In general, foxes spend much time above ground, lying up during the day in good cover. It is largely during winter and up to June before the cubs disperse that they occupy badger setts. For the rest of the year it is more often the vixen that spends the day below ground. When she emerges in the evening, usually before the badgers, there is no tentative scenting at the entrance. Instead, she usually comes out with a rush and quickly disappears into the undergrowth. Only when fox cubs are about do you see foxes for long near a sett.

It is rare for a fox to be seen emerging from the same hole as a badger. If this happens it is usually due to some emergency, such as when a fox has been hunted or frightened and takes refuge temporarily in an occupied hole. However, this is not always so. E. Bartlett told me of a sett in Surrey where, in early April, both fox and badger cubs emerged from the only exit. Both sets of cubs were estimated at about two months old as they kept near the entrance nosing at each other and tumbling at the slightest prod. The sett was dug out a month later and was found to consist of a main tunnel leading to a large cavern-like chamber in which the badger cubs had been brought up. From this chamber, two tunnels diverged. One was an 'upright chimney' leading to another badger chamber and the other led to the fox's lair which was very foul with debris. It is surprising that they tolerated each other's presence within a relatively small sett.

The rotting remains of prey, fox scent and urine are greatly disliked by badgers, and this causes them to use different parts of the sett or, if not breeding themselves, move off to another sett, leaving the vixen and her cubs in possession.

R.L. Willan (1963) describes how one year, badgers were using the whole of a 20-entrance sett; some rabbits were present, but no foxes. In the late autumn the badgers moved over to a smaller sett 70 m away, but returned to the main sett in February, when foxes had already moved in to one part. The badgers did not breed and put up with the large litter of fox cubs for some weeks, keeping to their part of the sett, but eventually they moved back again to the smaller sett nearby. However, in July, once the fox cubs had dispersed, the badgers thoroughly spring-cleaned the sett and took up residence once more.

I agree with Hancox (1973) that 'in general, badgers are dominant to foxes and can evict them at will, but this is not necessarily true at the breeding season'. In support of this, he cites an example in a wildlife park, in Avon, where a mixed community of badgers and foxes resulted from burrowing between adjacent pens. In this situation, the foxes were only dominant when they alone possessed cubs.

When badgers have cubs below, any intrusion by a fox usually meets with strong aggression. W. McGreggor watched a sow with a very small cub, probably up for the first time. It was poking out from below its mother at the sett entrance while she kept nosing it down the hole to keep it in. After collecting some bedding from close by, she disappeared down the entrance. Later a fox strolled nonchalantly over the sett area and when level with the occupied entrance, the sow badger shot out and nearly knocked it flying—chasing it off for 25 m before returning to the sett where the cub was.

Most other cases of this kind of aggression known to me occurred during the period February–May and involved sows which had small cubs. One instance occurred at 09.00 when Mrs H. Wilson was disturbed by pandemonium going on in the dell near her house in Worcestershire one morning. On going out she found a badger pushing a fox into a pool which was near the sett. The fox appeared to be trying to climb out of the water, but was kept from doing so by the badger. However, on seeing Mrs Wilson the badger dashed down the sett and the fox, very bedraggled, struggled out of the water, shook itself and ambled off across the fields.

Usually, these fights end with the fox running away, but occasionally this is not possible. There was such an incident on a cliff edge in North Wales which Abel James witnessed one morning. He was going to work at 07.00 and heard a scuffle on the cliffs above. He saw a vixen and realised she was on a ledge which had a dead end. He climbed up and saw that she was being followed along the ledge by a badger. Both animals were snarling at each other. The vixen, finding herself trapped, faced the badger and as a last bid for safety attempted to leap over her opponent, but the badger was too quick for her and caught her by one of the hind legs. For a few moments they rolled over and over on the ledge and then went over the side still locked together. When Abel James reached them they were both dead.

C.M. Cowland (1953) of Bovey Tracey, Devon, also recorded a mutual killing. He found in an oak wood the bodies of a medium-sized fox and a large badger locked together in death. The badger's jaws were firmly buried in the top of the fox's head. Presumably it was unable to withdraw its teeth and had died—possibly of suffocation. No other wounds were visible. Monica Edwards in her book *The Valley and the Farm* tells how she saw a badger take a vixen out of an outlying hole of the main sett and while still struggling in its jaws take it down another entrance.

Fights between foxes and badgers often attract attention because of the noise that ensues and the contestants are often too engrossed to notice a person's approach. Such a fight took place on 21 January near a sanatorium on the Quantocks, Somerset. When the noise was investigated, a fox and a badger were seen on the drive in furious combat and it was possible to get quite near before they were aware of being watched and rushed away.

Very occasionally a fox will kill a badger. Miss P.D. Hager (1957) describes a case of this in Chesham, Buckinghamshire, where the contestants were fighting in a lane, making a great deal of noise. When a car came along, the fox made off across a field and the badger was found to be dead. This was in July and could have involved a well-grown cub. No age was mentioned in the report.

Fights are not confined to adults. Brigadier Badcock heard a tremendous din coming from the rhododendrons near his house in Somerset. It went on for a good ten minutes. The contestants were found to be a badger cub and a fox cub. On being disturbed, the fox cub bolted and immediately the badger cub set off in pursuit. Badcock was astonished that two such small animals could have made such unearthly yells. This was in June when the cubs were only four or five months old.

There are many instances reported of adult badgers killing fox cubs

SETT TENANTS AND ASSOCIATIONS WITH OTHER ANIMALS

(Batten 1923). This can be the work of a boar, but is probably more often due to a sow with cubs. Most reports are from circumstantial evidence, but occasionally the deed has been witnessed. E. Clay described in *The Countryman* how he used to watch four half-grown fox cubs playing on the hillside opposite his home in Devon. One hot evening in June, he heard screams from that direction which continued during the time he was making his way towards the place. The screams came from a dense blackberry bush and were mingled with a grunting, snuffling noise. Parting the brambles he saw one of the fox cubs lying there, one back foot bitten off, its hind quarters apparently paralysed and skin torn on its shoulders. As it turned to look at him the head and fore quarters of a huge badger appeared from a hole at the back of the bush. It gripped the cub by the throat and dragged it swiftly down the hole.

Occasionally, it is the fox that is the aggressor and badger cubs the victims. Roger Burrows (1968) found vast quantities of badger hair in a fox dropping and on another occasion a large piece of badger cub skin. The badgers, of course, could have been eaten as carrion, but the circumstances did not suggest this.

Monica Edwards was watching a sett in Surrey in May, where there were two adult badgers, two yearlings and four cubs. A vixen was using one of the outer entrances. The badger cubs came out first, played and went in again. Then the boar emerged and went down the hole used by the vixen and stayed there for up to ten minutes. He emerged carrying a full-grown rabbit which he took down into the part of the sett where the family was living. In ten minutes, he came out again and went back to the fox's earth. This time he stayed down for more than 15 minutes. He then emerged carrying what Monica Edwards thought was the dead vixen as it seemed too small for the dog fox and too big for a cub. The boar took the dead fox down into the sett as he had done with the rabbit. Did the boar kill the vixen, or was it already dead when the badger arrived? It is frustrating not to know the answer, as the incident is a most intriguing one.

From these accounts of aggression between badgers and foxes it should not be assumed that fighting is usual. In the great majority of cases badgers and foxes are respectfully tolerant of each other's presence and go their own ways, but trespass by either on the other's territory when cubs are about may lead to aggression. So a wary armed neutrality might be a better way of expressing the relationship. Sometimes, however, all the rules are broken and Wijngaarden and Peppel (1964) cite a case of a sow badger adopting the cubs of a couple of foxes which had been shot and providing them with food.

There are certainly instances of badger cubs and fox cubs playing together very amicably. On the Wiltshire Downs, two badger cubs were watched playing for some time with a single fox cub in the early evening, in June, and Anthony Buxton wrote of an occasion when

> for about 2 hours till dusk in June, I had first 10 fox cubs and a vixen, and then 7 badgers all just underneath my tree. The badgers, of which 4 were cubs, played in one ring and the fox cubs in another, but one fox cub came across and sat down in the middle of the badger cubs, lolled his tongue out and 'laughed' at them.

However, they usually play independently.

INTERACTION WITH OTHER ANIMALS

Badgers often come into close association with other mammals during the course of their wanderings. Apart from their interaction with prey species, which will be described in Chapter 8, they may also come up against deer and various farm animals. Usually, badgers take very little notice of the larger mammals, but if the latter have young the reverse is not always so true.

Jim Rowbottom described to me how, when watching badgers, he saw a roe doe chase a sow and two cubs back to the sett when they approached too near. He assumed the doe had a kid nearby.

Similarly, Phil Drabble wrote to me about a boar badger which wandered in the direction of where he knew a roe kid was lying. When it got within about 10 m the doe rushed at the badger and passed over it at full speed, striking with her fore legs as she did so. The boar turned and immediately ran for the sett, but the doe returned to the attack twice before he got there. It was over an hour before he re-emerged!

Badgers commonly come across cattle. Sometimes the latter are very curious and sniff round the sett entrances at dusk, stamping and defecating. Disturbance of this kind may cause badgers to abandon a sett. However, when out in the fields badgers show less fear of cattle, but if they can avoid them when foraging, they will do so (p. 204).

This difference in behaviour when a badger is near to a sett and when some way away is very marked. It also applies to sheep. On the hillsides sheep and badgers take little notice of each other, but near a sett it is different. John Webster, who farms in Cumbria, told me of a sett in open fell-side which was heavily grazed by sheep. The bare area near the sett was very popular with sheep and more particularly with lambs, which lay out and played on it. If this area was occupied, emergence by the badgers was delayed for up to an hour. One occasion is of special interest and it is worth quoting from his excellent account:

> A lamb showed curiosity in the holes, scraping and putting his head into them. It then settled down with the ewe in the entrance of the main hole, thus inhibiting the emergence of the badgers which looked out from various other holes but did not emerge for some time. Suddenly, the sow, two cubs and two young adults came out one after the other and went a little distance up the fell where the sow started to collect some bedding. The cubs began to play and as their game rapidly became more wild they raced together towards the sett. This so excited the lamb that it too started galloping and leaping about as if to join in. The badgers, however, completely ignored it.
>
> Exactly a week later a lamb approached the sett, this time when two cubs were already out, some 50 m away. When they became aware of the lamb they immediately ran back to the sett and one slipped in through an outlying hole. The other, approaching the main hole, met the lamb directly. The lamb lowered its head and making slight butting motions caused the badger to stop and then run away on a wide detour to enter the sett by another hole well away on the other side. On both occasions the lamb showed interest in the badgers which in turn were very wary. When the ewe was present she was quite indifferent and apparently in no way apprehensive for the lamb's welfare.

SETT TENANTS AND ASSOCIATIONS WITH OTHER ANIMALS

Accounts of interactions with dogs vary considerably. Many badgers have been brought up in the company of dogs and the association has been companionable and amicable. Occasionally, a bitch has suckled a badger cub and helped bring it up. A most interesting example of this is given by Wickham Malins (1974) in his book, *Bully and the Badger*. He relates how his bull terrier bitch, which had never had puppies, was stimulated to produce milk by the badger cub and was able to suckle it successfully.

At the other end of the spectrum, hounds and various breeds of terriers will at once attack a badger if found above ground; hounds will probably kill it. Terriers, however, if they are experienced, seldom come to grips. Their normal role in the days of badger digging (now illegal in the UK) was to keep the badger at bay in the sett and prevent it from digging away from the men with the spades. This attitude was reflected in an incident told to me by Roger Muirhead concerning his five cairn terriers which lived in a state of armed neutrality with the badgers in a nearby wood. On returning from a walk, several hours before sunset, he heard noises which suggested that the dogs had cornered some animal. The four bitches soon came back but the dog stayed in the wood yapping. On investigation he found the dog had cornered a fully grown badger, but neither appeared to have attacked the other in earnest although they were making indecisive attacking runs at each other. He moved to pick up the terrier, but at that moment they came to grips and surprisingly it was the terrier that got a grip on the badger behind the jaw and above the eye. On picking up the dog this hold was eventually released and the badger ran off.

The most remarkable and vivid description I have read of an encounter between dogs and a badger was written by Fred Speakman (1965) in his book *A Forest by Night*. In this he eloquently describes how he witnessed a battle royal between an Alsatian and two smaller mongrel dogs on the one hand and a boar badger on the other. It occurred in a wood after dark. I cannot do better than quote extracts from Speakman's account as it typifies both the courage of the dogs and the stubborn, powerful defence of the badger.

> They are running and fighting together. Here and there, dodging and rushing, comes the short swift thud of a badger's feet, the high excited cries of dogs in chase. What a furious medley of sounds! Sharp cries of eagerness, agonised yelps of pain, high scream of badger, more yelping of a dog. The badger stands at bay. The face swings low to the ground, left and right in a move so swift no dog could avoid it. The badger is fighting this fight in his own way. He crouches to ground. He advances in short swift rushes. He retreats as quickly, swings about in his own length, faces the little dog at every turn. But the other is at his rear. He rushes in, bites and flashes away, safe, as the badger wheels to meet him. So they torment the strange beast, vicious feint and counterfeint, holding the badger's attention till one or the other can dash in. The Alsatian waits behind. Now she is there, full face to face, growling and barking in black anger, prancing lightly for all her weight, tail lifted high. The badger rushes forward, the Alsatian seizes it above the neck and throws it wide. The mongrels rush in. But the badger is on all fours again even before it reaches the ground. It has caught the Alsatian's lower jaw a ripping slash. It makes for the big dog again, the little dogs like fiends along its flanks. The Alsatian comes in at a charge. It flings

the badger high. A little dog is underneath as it lands. She cries high and terribly. The other undaunted dances near . . .

The badger is now low to the ground. His breadth is incredibly broad. He is so low the black belly is hidden under the body. The coarse grey hairs stand stiffly out like a bottle brush. His back is steeply arched . . . for all the world like a vast grey pear drop . . . there seems nothing for the dogs to bite on to. His defence is impenetrable, and that head of black and white is provided with jaws to whose snapping the gin trap is a toy. He swings to meet a dog. A cry rings out upon the night. The dog runs off limping, whimpering pitifully . . . The badger thuds off into the distance. The sounds grow quiet. He is gone to earth.

When Speakman found the dogs, the Alsatian had her head down, her jaw slashed and her flanks were heaving; beyond her were the two mongrels stretched out on their sides, both badly bitten. One of them subsequently died.

There have been a number of reported encounters between dogs and wild badgers which by contrast have been marked by curiosity and playfulness rather than aggression. One instance recorded by Donald Bradnam involved a six-month-old puppy, which when being taken out for a walk encountered a foraging badger. The animals played together quite happily while the dog's owner watched from a distance.

There seems little doubt that the great variation of behaviour shown by badgers and dogs when they meet is determined by the temperament and breed of the dog, and the past experience, or lack of it, on the part of both dog and badger.

6 The special senses

The size of the brain and the relative proportions of its parts usually reflect to some extent the degree of intelligence of its possessor and the relative importance of the various sense organs to the animal concerned.

A badger's brain, like that of most carnivores, is very well developed. This applies particularly to the two cerebral hemispheres which cover most of the dorsal region. Their surfaces are considerably convoluted, thus increasing the area of the cerebral cortex which is the main co-ordinating centre of the brain and the part one associates with the degree of intelligence.

The other parts of the brain which are particularly well developed are the olfactory lobes and the cerebellum. These are concerned with information sent by the nose and ears respectively. By contrast, the corpora quadrigemmina, which are completely hidden by the cerebral hemispheres, are small.

Figure 6.1 Brain of badger: dorsal view

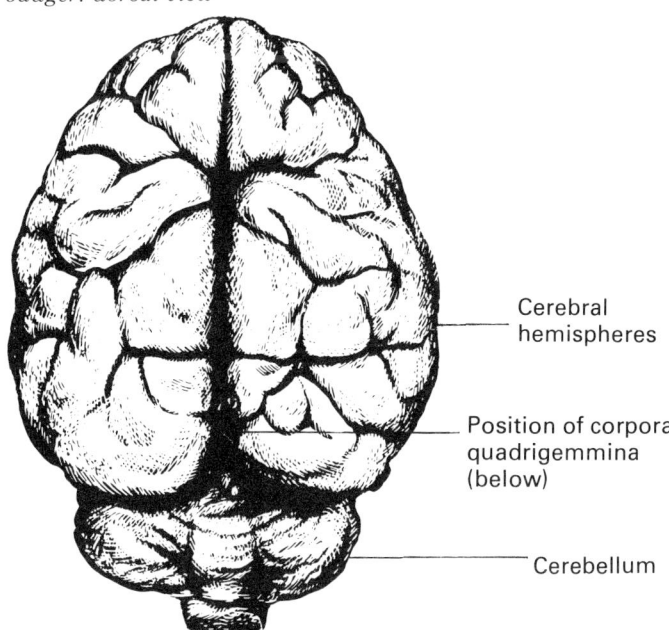

These are the structures which receive information from the eyes.

So, the gross structure of the brain suggests that the badger, like most members of the order *Carnivora*, is intelligent and that the senses of smell and hearing are well developed, but its eyesight is of less importance. This is largely confirmed by behavioural observations and experiments.

EYESIGHT

The eyes of most mammals which are adapted for good night vision have three main characteristics. The eyes are large and the pupil capable of much dilation so that as much light as possible can enter the pupil. The retina is very sensitive to low light intensities, which means that it has a high proportion of cells, called rods, compared with other cells, the cones, which are adapted for strong light conditions and may also be used in colour vision. There is present behind the retina a reflecting layer, the tapetum, which reflects back through the retina any light not absorbed on entry. In this way, the retinal cells are given a second chance to be stimulated so the sensitivity of the eye is greatly increased. It is the tapetum which gives the characteristic eye-reflection, often coloured, when a light is shone into the eyes of a nocturnal mammal.

If a badger's eyes fulfilled all these conditions, one would say that night sight was important to the animal. In fact, this is only partially true. A tapetum is certainly present, as most badger photographers know to their cost when they see on their photographs the blank, staring look caused by the reflection of their flash from this layer. It is also true that the retina consists largely of rods. There is no doubt that badgers have an aversion to strong sunlight. Wickham Malins (1974) writing about his tame badger 'Jess', said that 'in sunlight she was inclined to bury her head or shield her eyes with her paws and would sometimes retreat into a dark corner'. There are many instances recorded of this type of behaviour. But the eyes of badgers are unusually small compared with most mammals which have good night vision. So, these three characteristics suggest that night sight is probably quite good, but of less importance to the badger than the other main senses.

This conclusion is not surprising when one considers the very significant fact that badger cubs spend approximately the first eight weeks of their life underground in almost complete darkness and during this period have no opportunity of associating any event with sight stimuli. Even after they come above ground their learning processes are mainly associated with smells and sounds.

There are therefore two main questions to ask regarding badgers' eyesight—how well can they see and how important is sight to them? These questions are best answered by giving some illustrations.

I was watching a sett in June; there had been a severe drought; food for badgers was scarce and they were coming out unusually early. I was standing 5 m from the sett with the wind blowing steadily in my face. An adult sow emerged at 20.05, i.e. a full hour before sunset. I was wearing dull coloured clothes and kept quite still. She scented carefully, moved her head in most directions including mine, but was completely oblivious of my presence. Soon, two large cubs joined her and they played together, obviously not knowing that I was there, although to human eyes I was very plainly visible.

The factors in favour of my not being seen were: the light was very strong and in great contrast to the darkness below ground; the badgers were unsuspicious as they detected no human scent or sound; I made no movement; and I was not standing against a clear background.

John O'Connor described to me a rather different reaction to his presence.

> I reached the sett long before dark, and assuming that no badgers would be out for some time, made no attempt to conceal myself. I was standing on the edge of the sett when an adult badger backed into view with a load of earth. It stared at me in obvious surprise (I was about 3 m away), but when I made no sound or movement it returned to its digging. Within a few minutes it had reappeared three times with earth and stones and each time was obviously more curious. It would look at me several times and advance a few steps to try to scent me. Only after its fourth load did it become sufficiently curious to walk right up to me and get my scent. It then slowly retreated to the sett and did not re-emerge.

In this case the observer was seen at once and apparently recognised as something different from the usual surroundings of the sett. This caused the badger to show curiosity, but not fear as the senses of smell and hearing gave no alarm stimuli.

Paget and Middleton (1974b) described an occasion where the sett was in an open field and the observers about 50 m from it—a very long way! When a badger emerged, it looked straight at the watchers who were silhouetted above a low hedge and retreated hastily below. The wind was blowing strongly into the observers' faces and they had made no noise.

There are many other instances recorded where badgers have reacted at once to the sight of a strange silhouette in the neighbourhood of the sett, but in an open situation such as the one Paget described one suspects that badgers would show more caution when emerging and that vision might play a more important part than when the setts have greater cover nearby.

These examples strongly suggest that a badger's eyesight is more concerned with shapes than details; I think it very probable that a badger soon learns the outlines of the main objects in the vicinity of the sett and that any additional feature that is noted on emergence is treated with curiosity or suspicion until it has been investigated.

The reaction tends to vary according to what information the badger has received previously from its other important sense organs. If it is already slightly suspicious through smell or sound, the sight of a strange object will result in an alarm reaction, but if it is unsuspicious, curiosity is more likely to be shown.

The same seems to apply to movement. There is little doubt that badgers can detect movement quite easily even when unaccompanied by sound, but their reaction to it varies according to whether it is familiar or unfamiliar.

What happens when the light has nearly gone? There is plenty of evidence that a silhouette against the night sky will cause alarm in just the same way as in better light. In addition, anything which reflects any remaining light, such as the shiny surface of a photographic flashgun, the moving hands of an observer or a white handkerchief or article of clothing, is often noticed at once on emergence and treated in the same way as any other unfamiliar object when seen in better light. My impression is that their eyesight is good in low intensity light and certainly good enough to recognise their own kind very quickly by the characteristic white pattern of the head.

To sum up: because a badger's retina consists largely of rods it tends to be blinded by strong daylight—especially so on first emerging from the darkness of its tunnel; also it does not have the visual acuity which enables it to see small details from any distance, but shapes, silhouettes and movement can easily be detected even under extremely poor light conditions. I believe the importance of eyesight to a badger is to alert the animal in a general way to possible danger so that its other sense organs may analyse the situation in more detail, or confirm a suspicion that has already been aroused.

I have already discussed a badger's reaction to normal torchlight. The fact that they usually take no notice of it, so long as it is directed steadily from a position rather above them, does not mean that they do not see it. On the contrary, if they have previously become suspicious, then the light can be the last straw and they may react in alarm.

Badgers seldom react at all to a torch with a red filter, but it is possible that they merely fail to react because the light intensity is low.

Badgers have often been described as shortsighted, but this would appear to apply much more to cubs than adults. Paget's evidence, previously described, suggests that an adult can see a silhouette at 50 m. However, cubs are very different. The time when the cubs first open their eyes is variable, although five weeks is probably the usual, but it is unlikely that they can focus them properly until about seven weeks at the earliest. By 8–9 weeks they will follow an object with their eyes if it moves within 300 mm of their head. As they get older the range of focus increases. Most badger watchers have had experience of young cubs coming right up to them if the wind is favourable without apparently seeing them. It is when the watcher is scented that the cubs show alarm. It is almost impossible in these cases to be certain whether this lack of fear is due to the inability to see or to lack of experience, but in cubs of 10–11 weeks old the former would appear to be one reason.

HEARING

I know of no critical experiments done on the range of sounds detected by badgers. For those carnivores which have been investigated, the useful upper limit falls well above that of man. Observations on badgers suggest that they are no exception to this generalisation. B. Vesey Fitzgerald wrote that badgers can hear a Galton whistle, and I know from my own experience that they can hear from 6 m the very high-pitched whine of a re-charging electronic flash unit. This suggests a good sensitivity to high frequencies.

Many carnivores which feed on small prey can hear the high-pitched squeaks (some of which are ultrasonic to us) of small rodents. It is likely that badgers can do this too, but it is extremely difficult to know from a badger's behaviour whether the prey is detected by sound or scent. For example, earthworms can be detected when still below ground; can they hear the scraping of their chaetae or can they smell them? Perhaps it's a combination of both.

Badgers make a great variety of sounds which are of value for communication (p. 140). These cover a wide range of frequencies from the low murmuring of the sow, when with young cubs, to the high-pitched whickering of the cubs and the staccato warning notes of the sow. Within this range, a badger's hearing seems to be quite as good as ours and is probably better in the higher frequencies.

Low frequency vibrations may be felt directly via the ground. There is no doubt that badgers, when crossing roads, can detect oncoming traffic in this way. Unfortunately for them, their response to this is similar to a hedgehog's, which is to stop still. In consequence, large numbers are killed on the roads.

When filming badgers at night with Professor Hewer (1954) the sound of the camera was a considerable problem. Although the camera was encased in a box with a plate glass front and heavily blimped (lined with rubber, etc. to reduce noise) the badgers heard the motor mechanism without difficulty. Our first success came when it was raining heavily, so we assumed that the noise of the rain had effectively drowned that of the camera. So we fixed up a device which kept up a continuous low frequency vibration, which we hoped would have the same effect, but this did not succeed as the badgers appeared to be able to distinguish both sounds at the same time. We finally relied on their becoming habituated to the actual sound of the camera.

Badgers quickly become habituated to noises if repeated often enough. In built-up areas they soon get used to the sound of traffic going by, gates banging and dogs barking. Badgers which made their sett in a railway embankment on the main London–Hastings line seemed to take no notice of the vibration and noise whenever a train passed above them. A.D. Mennear watched badgers at a sett on a flight path to London's Heathrow airport. Low flying jet aircraft had no effect on the badger's activities, though men with a tractor 200 m away made them pause and listen. John Whall made use of this tolerance to sound once they were accustomed to it, by taking a transistor radio with him when watching. They took little notice of either talking or music. Perhaps they enjoyed it!

The response of badgers to sudden sounds is often immediate and striking. It is the unexpected nature of the sound that mainly causes alarm. This is particularly obvious around the time of emergence when the badgers are specially alert to possible danger. The sudden crackle of leaves, the noise of a camera shutter or the rustle of clothing often result in a dash for home. The fear reaction to a sudden noise appears to be inherited, but gradually a cub learns to discriminate and associate certain events with these sounds and less notice is taken of some of them. However, with cubs of up to about 12 weeks any sudden sound may cause temporary alarm. They may be startled by the thumping of a rabbit nearby or a bird flying noisily to its roost, even some dry leaves blown by a gust of wind may cause them to bolt. Distant sounds seldom disturb badgers. You often see a badger pause on emerging when it hears the sound of a dog barking or distant human voices. It seldom retreats unless the sounds get nearer.

Hewer and I were able to film badgers quite successfully at Camberley in 1953, on Coronation night, when fireworks were being let off all around us, although none was nearer than 100 m.

Brigadier Fryer gave me an even more extreme example of this tolerance to distant sounds. He regularly watched at a sett on War Department ground and on one occasion when troops were camping out during night exercises, even the resulting rifle fire (blanks) and thunder flashes didn't seem to worry the badgers for long. They stopped and listened, but did not always bother to seek refuge in their holes. In contrast, even a slight sudden noise near the sett would alarm them.

THE SPECIAL SENSES

Once badgers are out and foraging certain shuffling sounds may cause them to show curiosity. John Brodribb recounted how, when watching had been uninteresting, he had moved about quite noisily and badgers had come to investigate the sounds he made. One night, he was trying to move quietly, imitating the noise of a foraging badger, when he slipped and landed with a terrific crash. At once, two badgers from different directions rushed towards him, one coming within 3 m of where he was lying!

It is certainly possible to get quite near badgers by imitating their own movements. Also, by making scratching noises similar to a badger grooming, you can often calm them down if they are agitated.

SMELL

The sense of smell in badgers is extraordinarily well developed, and is undoubtedly their most important sense. This is not surprising, considering the great development of the scroll bones in the nasal chambers. These provide an extremely large surface area for the sensory epithelium which covers them. Some of this area is more concerned with temperature and humidity regulation of the inhaled air, but nevertheless, the olfactory portion is extensive and is provided with a generous supply of nerve endings.

It is probable that by increasing the humidity of the air as it passes through the first part of the nasal chambers, the sensitivity of the olfactory region further back is increased.

A badger's world is a world of smells. For us, when sight and hearing are the dominant senses and our olfactory sense is so poorly developed, it is difficult to imagine what it is like to live in a world dominated by smells. However, for badgers scent plays a leading role in recognising individuals and their sexual state, for finding food, detecting danger and finding their way about.

When the badger emerges from its sett in the dusk, a very sensitive snout is enquiringly raised to get the evening's news. First in one direction, then in another, it tests the air. In this way it detects the slightest sign of danger.

Figure 6.2 Badgers scent the air cautiously when emerging from the sett

If a badger catches the strong scent of the watcher, it will withdraw immediately; if not quite sure and only slightly suspicious, it may move its head up and down, testing the air at different levels and sniffing all the time as it does so. If no scent of man is detected, out it will come and soon it will relax.

On one occasion, a sow had come out quite unsuspicious of my presence, and her cubs, which were quite small, soon followed and started to fiddle about with a piece of stick near the entrance. But the wind was variable and the sow suddenly got my scent. There was a gruff, suppressed but urgent, bark of warning and she bundled the cubs down the hole in front of her. Experiences such as this cause cubs to associate human scent with danger.

When badgers are foraging, they keep their noses very near the ground and as they search amongst the vegetation they constantly make loud snuffling noises. These are made in two ways and serve different purposes. The first is caused by a rapid intake of breath through the nose; this helps to concentrate faint scents and enables the badger to locate prey more easily. It occurs in dramatic fashion when a badger is suddenly frightened—the snort of fear probably helping the animal to obtain more olfactory information about the cause of alarm. Other snuffling noises are made when air is blown out through the nostrils to clear away particles taken in when foraging under dusty conditions.

Badgers can be very pre-occupied when foraging, but periodically they will raise their snouts and scent the air carefully. When the snout is near the ground other smells would tend to overcome the direct wind-borne scents. This would explain why a travelling badger may come quite near before detecting your presence if you stand still, even if the wind is wrong. Under different circumstances, such as when the badger is emerging from its sett, it would detect you at once even from a much greater distance away.

The acuteness of a badger's sense of smell is illustrated by experiments done by Howard Lancum (1954). At 11.00 in late May, he placed the palm of his hand on a badger path for one minute. On watching the sett that evening, two adults emerged together at 22.00 followed by three cubs. When the boar reached the spot, it stopped, sniffed and made a slight detour before continuing, but the sow would not pass and returned with the cubs to the sett. Two days later at a different sett he repeated the experiment—this time at 15.30. At 22.15 that evening an adult emerged and on reaching the spot, 'shied like a frightened horse and bolted underground'. Two hours later it had not re-appeared.

Eric Ashby, some years later, repeated the experiment in the New Forest, using gum boots instead of the palm of the hand. He successfully filmed the startled reaction of a badger when it reached the exact place.

Their power of detecting their own scent trails is even more remarkable. There was a well-marked trail across a grassy field from a wood in which there was a sett. The field was ploughed up and put down as corn, but the badgers continued to use exactly the same route, presumably by the scent that still lingered, even though all visual signs of the path had gone.

There would appear to be a level of scent detection below which a badger does not react. It is common experience when watching badgers that on some evenings they react to your scent more obviously than on other occasions, irrespective of the direction of the wind. This greater sensitivity usually seems to be due to an increase in humidity of the atmosphere—just as everything to us seems to smell more after rain. It is possible, therefore, as Frazer Darling found for red deer, that on these occasions a badger may react in panic to your scent even if you are some way off because you are scented with the same intensity as when you are nearer on a drier night.

Temperature also seems to be an important factor in scent detection. Budgett (1933) working on dogs, showed that the optimal condition for tracking was when the ground temperature was a little higher than the air temperature. Because air temperature falls faster than ground temperature in the evening, the latter is usually the best time for tracking and is in fact the time so often used by carnivores when hunting. This would also apply to the badger.

At the breeding season sows appear to be more sensitive to human scent than boars, a point corroborated by Howard Lancum's experiment quoted earlier. It seems that at the breeding season the level of scent stimuli at which the sow will react is lowered.

The acuteness of a badger's sense of smell is very great but it is probably its ability to discriminate between scents that makes it so outstanding. Badgers can easily detect differences in human scent. This is very evident with tame badgers when a stranger comes into the room. It is also obvious in the wild where badgers have become habituated to the scent of a particular watcher. Tony Bennett regularly visited a sett near Bath, and by scattering sultanas among the leaves at each visit was able to bring the cubs right up to him even when the wind was wrong; but each time a stranger accompanied him they were much more suspicious and only their love of sultanas would eventually persuade them to come near.

Badgers can also detect differences between the scents of other badgers, both members of their own social group and of neighbouring social groups (p. 142).

This power to discriminate between a wide variety of subtle scents obviously has a bearing too on a badger's ability to find many kinds of food; it also enables it to find its way about.

One can imagine how a badger gradually learns to orientate itself when on its travels by the scent pictures it has previously experienced and memorised of specific parts of its home range. This is clearly reflected in the development of exploratory behaviour in cubs. At first, they keep to their underground system, then to the area just round the entrance and later on they explore the whole area around the sett. However, it is some time before they have built up sufficient experience of the terrain (I believe, mainly by smell), that they venture far from home. It is then often necessary to provide their own scent trails for finding their way back.

7 Activity patterns

Badgers are largely nocturnal or crepuscular, although daylight appearances are by no means unusual in secluded places. Typically, adult badgers emerge from their setts between sunset and darkness during the summer months, but only after dark from October to March.

FACTORS AFFECTING EMERGENCE TIMES

Those who have watched badgers regularly at a particular sett can often predict the time of a first emergence with remarkable accuracy, especially during spring and autumn, but there are many exceptional evenings when the watcher is widely out in his forecast. Badgers, like humans, are sometimes fickle over keeping appointments!

By noting the time of first emergences throughout the year it is possible to produce an 'average emergence time' graph which for most evenings from March to November would enable the badger watcher to predict emergence with an accuracy of about plus or minus 15 minutes.

I published such a graph (1948) for a Cotswold site and more recently others have done so for various parts of Britain and Europe. But I wish to refer specially to a comprehensive four-year study by Keith Neal and Roger Avery (1956) because all data relate to the same sett and any variation from year to year will be correlated with changes in badger population and weather conditions.

The sett was in a relatively undisturbed area of the Quantock Hills, Somerset. Observations were made from a tree where possible disturbance through scent was highly unlikely. All the main holes were visible from the tree. Occasions when the badgers appeared to be suspicious were not included in the data (400 observations). Emergence was defined as the first time a badger came completely out of an entrance. The graph, Figure 7.1, shows the results to which I have added data sent to me by F. Vaughan who watched a rather similar sett near Yeovil, some 56 km away. These additional times were most helpful as they increased the number of observations made during the difficult watching months of December–February.

Certain generalisations may be made from this graph. First, the curve showing the average time of emergence for each month of the year is a smooth one which runs very roughly parallel with sunset times between

Figure 7.1 Times (GMT) of first emergences when undisturbed (432 observations at two setts in Somerset by K.R.C. Neal, R.A. Avery and F. Vaughan). a) Throughout the year; b) during winter

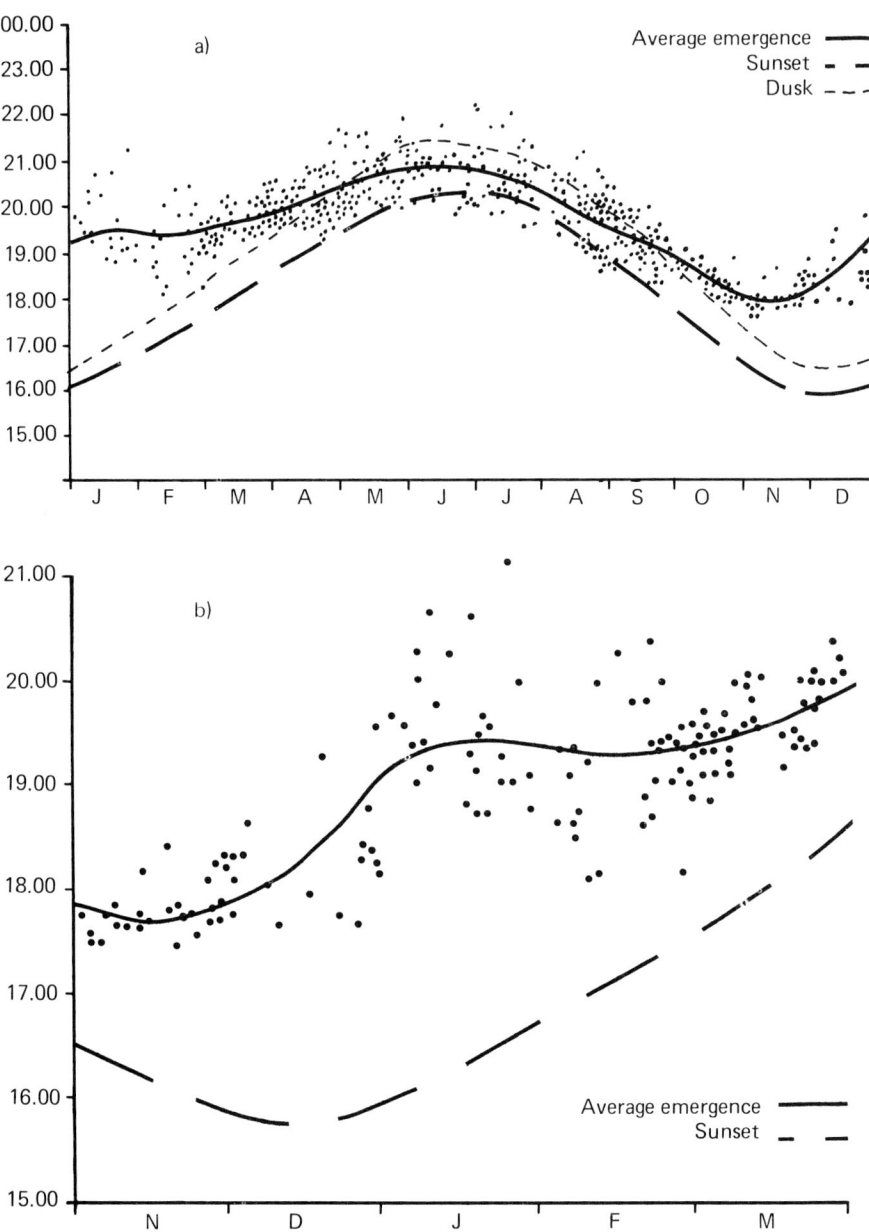

March and September, except for a flattening during midsummer. This suggests that emergence is to some extent determined by seasonal factors. Secondly, there is much variation during every month, but more so at certain seasons. So one must infer that there are other factors which influence the general pattern of emergence from one night to another.

One can also derive from the graph certain seasonal characteristics. In January and February, emergence took place after dark and was largely unpredictable as there was as much as three hours' difference from one night to another. In March, emergence was much more regular and occurred soon after dark. In April, a very similar pattern was shown, but towards the end of the month some emergences were well before dark, while others were much later. From May to August, the majority of first sightings were before dark, many of them in excellent light and occasionally even before the sun had set. However, in July and August especially, there was a much wider range of times with occasional very late emergences as well as some particularly early ones. In September, times were very regular and tended to be at late dusk. In October and November, emergence also followed a regular pattern but it occurred nearly always after dark—in November usually at least an hour after. In December, times were unpredictable, and on a few nights no badgers emerged at all.

It is evident that the factors influencing emergence times are complex and often inter-related; these include light intensity, the amount of cover round the sett, disturbance by people or animals, the weather, the amount of available food and the social composition of the sett residents.

Light Intensity

I previously assumed (Neal 1948) that light intensity was the main factor which determined the time of emergence, and this conclusion has been generally accepted by other observers on the grounds that the curve showing average times of emergence between March and October runs roughly parallel with sunset times. I also assumed that local or seasonal variation in this pattern was caused by other factors, such as shorter nights, difficulty in getting sufficient food, weather conditions, etc. This hypothesis sounds very plausible, but if true, it must be reconciled with certain other facts.

First, it is evident that light intensity cannot be a factor during the period November–March, when the first badger comes out well after dark under normal circumstances. Second, it is obvious that not all the resident badgers come out at the same time, so they do so under varying conditions, including those of light intensity. For example, on the night of 18-19 April 1955, during an all-night watch, Neal and Avery saw the first adult emerge at 20.20, at late dusk, but seven other resident adults or yearlings came out at the following times: 21.55, 22.15, 00.15 (2), 00.22 (2) and 00.55! This was certainly exceptional, but on a great many other occasions the emergence of adults was spread over a period of well over an hour. Those badgers which emerged later were obviously not affected by light intensity.

One could argue that if light intensity were the main factor in summer, emergence times would vary in different geographical localities according to sunset times, although one would expect this to be true only as a generalisation as other factors such as disturbance or the availability of food would modify them. On the whole, this expectation is not fulfilled.

If you compare the results in Somerset with those in Essex, you would expect, because of differences in sunset times, that the Essex emergences would be on average about 15 minutes earlier. But data provided by W.W. Page (over 700 observations by a number of observers over several years, entailing over 40 different setts) show a graph very similar to that of

Somerset for most of the year but from April–October emergences are all later by about 15 minutes (rather less in June). So in terms of sunset times, this would be about half-an-hour later—not what you would expect if light intensity were the main factor. Evidently, other factors of greater importance are involved.

Similarly, you would expect that as you go north, emergence times in summer would be later due to the later sunset times. Mary Marsh sent me her results from regular watching throughout the year at a sett in Scotland, in Inverness-shire. These showed that, compared with the south, emergence was earlier during the period May–August, in June by about an hour. Again, this result cannot be related to sunset times and suggests that other more important factors, such as food availability are operating. Skoog's data from Sweden suggests the same thing, as emergence there in July is often as early as 19.00.

With all this evidence against the light intensity theory, let us now consider evidence in its favour.

Examination of any emergence graph, whatever the geographical location, shows that the greatest correlation between light intensity and emergence time is in spring and autumn when emergence is usually at dusk and is very consistent. If you watch a particular sett at these times of the year, variation from one evening to another can often be related to other factors which indirectly affect light intensity, such as cloud cover or moonlight. Cloud cover often causes an earlier emergence. This was also commented upon by Lloyd (1968) who considered it to be the most important factor accounting for the difference in emergence times between eastern and western regions of southern England.

Moonlight, especially when it falls directly on to a sett entrance, has also been given as an important local factor in delaying emergence. I have certainly found this to be true in March and April at setts where there has been little ground cover, but I have no evidence that it makes much difference in the autumn—possibly because at this time the trees are in full leaf. Norah Burke (1964) noted that a favourite exit at one sett was not used on particular nights when moonlight shone down it.

It has often been noticed that the emergence times at different setts in the same neighbourhood may be consistently different by as much as 30 minutes. In some of these instances, light intensity appears to be the most probable reason. Michael Harrison studied two setts on a hill in Somerset, near Yeovil, one being on the top and the other on the north-facing slope. Both were large setts with many entrances. The sett on the top of the hill received the rays from the setting sun, the other did not and emergence from the hilltop sett was consistently about half an hour later than from the other. Harrison also mentioned that although both setts had cover from a number of elder trees, there was a significant difference in emergence times during spring 1957, compared with the previous and subsequent years: in 1957 the badgers came out appreciably earlier. This was an extremely mild spring and elders were in leaf by early April, but in 1958 they had not reached a comparable stage of leafage until about the end of May. In addition, he points out that after full leafage was on the trees, the records for all three years merged indiscriminately.

These instances provide good circumstantial evidence that light intensity plays an important part, especially in spring. So we have apparently conflicting evidence for the importance of this factor.

Cover

I have already alluded to tree and cloud cover as factors which may affect light intensity and therefore the time of emergence. I now want to consider the vegetational cover near a sett which allows the badgers to emerge and go off unseen. In a previous chapter I commented upon the importance of this factor in the choice of location for a sett. I believe this allows earlier emergence especially at certain times of the year.

D.R. Scott (1960) made an interesting comparison between the emergence times from three setts in south-west Essex which showed considerable, but consistent, differences between March and August. Calling the setts A, B and C, he found that the latest emergences were at A and the earliest at C, B being intermediate. A was in high beech forest without much undergrowth and with a carpet of dead leaves, the nearest ground cover being bracken some 50 m from the sett. Here, there was little general disturbance except the noise of distant traffic but the sett was close to several well-used footpaths. B was on private land and not subjected to human disturbance at all in the evening; cover was better than at A as there were a few bushes near some of the sett entrances and the ground was grassy. There was less background noise than at A. Sett C, where emergence was earliest, was in forest land, traffic noise was considerable and a footpath used by many people in the summer months ran alongside the actual sett. However, cover around the sett was excellent consisting of holly bushes and brambles which extended some distance. Scott concluded that good cover was by far the most important factor; when that was present badgers would tolerate noise and human disturbance and quickly slip away after emerging, lying low temporarily if people passed near.

George Barker also came to a similar conclusion about six setts at Old Winchester Hill, Hampshire. The average times of emergence during August and early September were 20.12, 20.32 and 20.35 at setts where cover was good; 21.10 and 21.20 where there was more disturbance and 22.15 at a sett in a yew wood where there was a total lack of ground cover and which was also subjected to considerable human disturbance.

Disturbance

Disturbance is a term which is difficult to define as it includes many forms of interference. At a sett where disturbance of any kind is minimal, badgers are likely to emerge earlier in the summer than at comparable setts which are disturbed. So it is wise when watching at very isolated and undisturbed setts in summer to get there well before the 'expected' time of emergence. 'Inexplicable' blank nights are often due to the watcher getting there after the badgers have gone off!

N.B. Palmer (1959) made continuous records over a period of nine months at two setts in the Cheltenham district. Both were in similar types of deciduous woodland, but whereas A was subjected to regular blocking by the hunt in winter and spring, pigeon shooting, children playing in the wood, especially in summer and occasional gassing or shooting of badgers, B showed no sign of human disturbance. Palmer found that during the three months' period when sett A was relatively undisturbed, emergence was more comparable with that of B, but at other times it was $1-1\frac{1}{2}$ hours later. During one week's intensive watching at sett A during September, he showed that emergence was much later on the Saturday and Sunday evenings—when children played round the sett during the day.

Palmer also showed that behaviour varied considerably at the two setts. At A the badgers went off quickly and started foraging almost at once; they were also very quiet and very little play occurred near the sett. At B the badgers stayed close to the sett for up to an hour, there was much play between the cubs and considerably more vocalisation.

Regular disturbance caused by background noise seems to have little effect as badgers become habituated to traffic and human voices quite quickly. However, any unusual or sudden disturbance has a considerable effect, especially if it involves the sett itself. Trampling by cattle over the sett will cause late emergence and possibly temporary abandonment, and the blocking of a sett by the hunt will cause very late emergence on subsequent nights unless the blocking is very light and the material used quickly removed.

Late emergence on a particular night is of course often due to suspicion caused by human scent, unusual sounds or the proximity of dogs or other domestic animals.

Weather

I have already discussed the effect of cloud cover and moonlight, but other weather factors are also important. Badgers appear to dislike emerging into heavy rain, although once out foraging they seem to take little notice of it. Görgen Göransson (1974) in a detailed study of emergence times and activity periods showed that during September, in Sweden, emergence times were extremely regular night after night except for one night when there was heavy rain associated with a drop in temperature and the badgers emerged $1\frac{1}{2}$ hours later. In general, slight rain and wet conditions after rain have little effect. Wet soil when temperatures are not too low are ideal for foraging for earthworms and this may bring them out earlier.

Windy nights may delay emergence. Badgers are easily frightened by strong winds and may often panic if there is a sudden gust with attendant noises. It is likely that under very windy conditions, the sense of smell becomes unreliable and the animal is nervous in consequence.

There is little evidence that temperature has much effect on emergence, except perhaps indirectly by affecting feeding conditions. In the north, where winters are more severe, emergence during the colder months may be earlier than in the south, possibly because some food (particularly earthworms) is more easily obtained before the temperature drops too much.

Badgers will emerge under extremely cold conditions. The late James Fisher told me how he tracked badgers in the snow in Northamptonshire after a night when the temperature fell to $-18°C$ $(-2°F)$! Badgers will often come out when snow is on the ground and have been seen to emerge not appreciably later than usual when snow was actually falling. Donald Bradnam recorded one emergence during a snow blizzard in February 1969, and Graham Moysey (1959) recorded on an automatic device the emergence of two badgers at 03.00 at a Devon sett following a fall of snow; however, they only remained outside for 30 minutes. By contrast, on the same night at a nearby sett, one badger emerged at 19.00 and the other eight hours later. So there is much variation in reaction to snowy conditions.

Figure 7.2 Returning after a forage in the snow

Availability of Food

This is one of the most important factors which affect emergence. Its effect is particularly evident in summer when short nights limit foraging time. When there is plenty of food about, as in a wet season, emergence may be near to the expected time, but given a sustained period of drought in June when the earth is hard and cracked and the earthworms impossible to reach, the badgers will be out much earlier. During the hot, dry summer of 1975, at a particular sett in the Brendon Hills in June, the badgers were out by 19.30 night after night. This was a full hour earlier than the average for that time of the year. Similar behaviour was reported from many other parts of the country during this period of drought.

These early emergences in summer also take place commonly in regions where food is comparatively scarce every year during particular months. This is so in Scandinavia and the Highlands of Scotland, where emergence in June may be as early as 19.15.

Periods of glut may have the opposite effect, so that after a night of abundance, a particular badger may not emerge the following night until very late—sometimes not at all.

Social Composition of the Sett Residents

This is another factor which modifies emergence times considerably at certain periods of the year. It applies particularly to the presence or absence of cubs. Neal and Avery (1956) found that after the cubs were about a fortnight old, it was almost invariably the sow which emerged first. This was also true of March and early April, when she came out particularly early, presumably to find enough food to keep up her milk supply. However, once the cubs had been above ground for about two weeks, they were the first to emerge and this was so for most of the summer months. It was curious that one cub regularly emerged much earlier than the others, but when all had come out, the sow usually followed quite quickly. There seems little doubt that when a family is large, the cubs stimulate each other to come out earlier, play being a compelling reason for emergence.

For setts where there are only two or three adults or yearlings, and no cubs, emergence is often much later during the summer; it can be well after dark. This may be one of the causes of differences in emergence times from one year to another at a particular sett.

Figure 7.3 Cautious emergence at dusk: three adults and a 12-week-old cub

General Conclusions

From the foregoing account, it is clear that a number of factors, many of them inter-related, influence the time of emergence. Some factors stimulate a badger to come above ground, the predominant one for most of the year being hunger, but also social activities such as mating, playing, digging and bedding collecting. Other factors retard emergence; disturbance, lack of cover, strong light intensity and certain weather conditions. I believe the approximate time of emergence is primarily determined by the animal's biological clock. This is a physiological mechanism which roughly determines the periodicity of such functions as sleep and activity; hunger being an important component in this mechanism. But variations of this generalised pattern are due to differences in the strength of the factors advancing or retarding emergence. In other words, the time of emergence is the resultant of opposing forces on any particular night.

If you consider the problem from this point of view, most inconsistencies and variations can be explained. Referring once more to the graph (Figure 7.1), the random nature of emergence in January is probably due to several factors. There is less urgency to feed as the badgers have considerable reserves of fat and often the weather limits the opportunities for finding food. This may be due to a succession of frosty nights, or to a sudden drop in temperature during a particular night. But, towards the end of the month activity is more pronounced due to the onset of the breeding season.

During February, with the birth of the cubs, there is greater need for the sow to feed, but she does not leave her cubs for long periods. However, the adult boars may be very active as this is the start of the mating season and territorial activity is considerable. These factors probably account for the earlier emergences during this month.

March is a very busy month for sows with cubs. There is an urgency to find enough food for adequate lactation and much time is spent on sett excavation and bedding collection. These factors exert pressure on the sow to come out as early as possible, but other factors deter her from doing so, at any rate before dark, as at this time of year there is far less cover available and darkness becomes an essential substitute. So light intensity becomes the governing factor in determining her emergence and she comes out as soon as it is dark enough. This explains why first emergence times during March are so consistent. However, adults other than lactating sows, may not be in such a hurry to emerge. For example, Neal and Avery found that while the sow emerged regularly throughout the month around dusk, the next adult to come out on some evenings was more than an hour later and others came out later still.

This pattern of behaviour continues into April, but by the end of the month, and into May, more cover is available due to the growth of vegetation and by then the cubs are regularly above ground. With cubs becoming bigger, play is now an important reason for early emergence, and if there is little disturbance, and cover near the sett, it will take place before dark.

By June, the cubs are weaned and more independent. Once more the main driving force is food. Only disturbance will prevent an early emergence as there is usually plenty of cover to counteract the disadvantages of being out in daylight. If food is scarce, emergence can be extremely early.

In July and August, the need for food by all members of the social group is paramount. The cubs are growing fast and the adults have the opportunity to make up for their loss in weight during the breeding and mating season.

From September–mid November the main pre-occupation is again feeding, as food is usually abundant and badgers put on most of their fat in preparation for the leaner period of winter. At this time the badgers may forage steadily throughout the night. As feeding is the main pursuit for all members of the group, emergence times are very consistent once more. Neal and Avery found that in September, seven badgers emerged each night within a period ranging from 27–56 minutes of each other. As the season advances and the nights get longer there are more hours of darkness in which to feed, so emergence, usually well before dark in September, becomes after dark in October and begins to be even later in November.

By the second half of November and during December, adults emerge very irregularly, although cubs of the year may continue to come out soon after dark, especially if the nights are damp and mild. There is usually a greater need for the cubs to put on more fat, especially if food was scarce in the summer or autumn. December is the most dormant period of the year for the adults. Occasionally, they may be out for several hours, but on some nights they may not emerge at all and on others, only for a few minutes.

So in my opinion it is the badger's internal biological clock that sets the general pattern throughout the year, the necessity for obtaining enough food according to seasonal and daily requirements (which differ markedly in individual badgers) that regulates the time setting, and that other factors such as light intensity, disturbance and lack of cover on the one hand and the urge to play, collect bedding, mate or mark out territory on the other, modify the time slightly from evening to evening.

TIMES OF RETURN

The results of 42 watches in Somerset are given in Figure 7.4. Each dot represents the return of a single badger and illustrates the independent nature of individuals on their nightly activities. On one occasion in September seven badgers returned over a period of two hours.

The pattern shows similar characteristics to that of emergence, but in an inverse manner. Thus, returns during the winter were well before daylight, but in the period April–September they were mainly spread out during the hour and a half before sunrise. Occasionally returns were long after dawn; this applied particularly to cubs in August and September.

THE PERIOD OF ACTIVITY

When emergence and return times are plotted on the same graph, you find that the active period varies seasonally from an average of $6\frac{1}{2}$ hours in midsummer, to over 11 hours in November. But, such a graph takes no account of what happens in the interim period between emergence and return. It is important to know whether periodic returns to the sett are made throughout the night, whether the behaviour pattern changes throughout the year and how much time is spent in such activities near the sett as digging, bedding collecting, grooming and play compared with foraging well away from the sett.

Graham Moysey (1959) in Devon and later George Barker at Old Winchester Hill were able to record the comings and goings of badgers by means of automatic recording devices. More recently Görgen Göransson (1974), working in Sweden with more sophisticated apparatus, was able to record

Figure 7.4 Times of return to sett in the early morning. From A.M. Milburn (Wiltshire and Worcestershire) and K.R.C. Neal and R.A. Avery (Somerset). Each vertical row of dots represents the return times of the various resident badgers

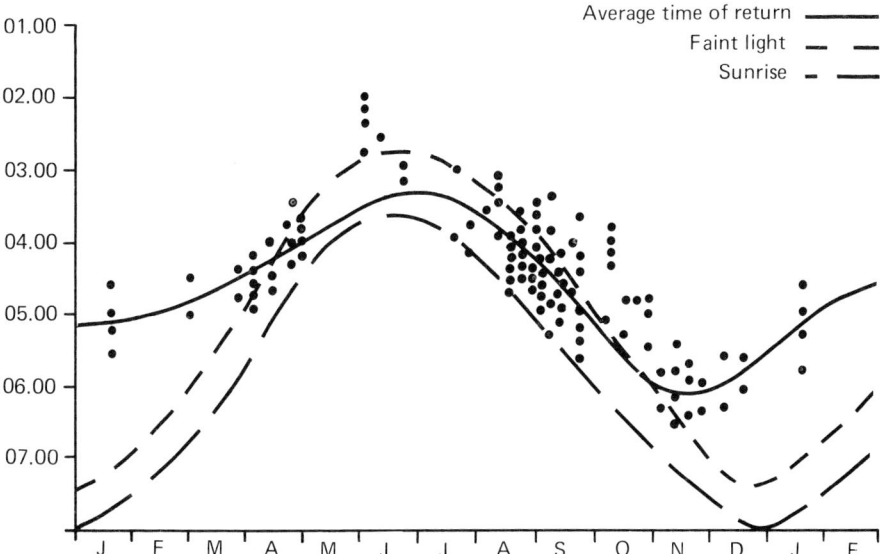

the time of each emergence and return and distinguish the direction of travel. This allowed him to calculate the number of badgers active each night (six), the total activity period and the time interval when no animals were in the sett.

It is now possible to fill in some of the details of night activity throughout the year.

In winter, Barker showed that badgers often did not emerge until after midnight but at other times after an early emergence, returned quite soon. Under extreme weather conditions they were sometimes only out for half-an-hour, and occasionally no badger emerged at all. During one cold spell, when snow was on the ground, the badgers spent several consecutive nights below and when they did emerge, did not go far. Exceptionally, a single animal travelled 3–4 km in the snow.

The time spent by badgers near the sett varies considerably according to season. This period usually includes a time following emergence when there is often grooming, play and general exploratory behaviour, and a similar period on return before they retire for the day. However, at certain seasons other activities such as digging, mating and those concerned with the presence of cubs may cause them to be around the sett area much more, and return visits during the night may occur.

Most time is spent near the sett from mid-January until mid-May, with a peak in February. This is the main period of mating, cub development and social activity. In February, in particular, there is often much rutting activity in the sett area. Also for the first few weeks after giving birth, the sow usually waits around the sett until other members of the social group have gone off before leaving herself for a short spell of foraging, and when the cubs are older, but still confined below ground, she makes periodic returns to the sett. During ten consecutive nights in early April when

Hewer and I were filming, the sow usually made four short sorties each night in order to forage.

Once the cubs are above ground, much time is spent in play and exploratory behaviour around the sett area and from May–July some members of the social group usually remain near the sett for some time.

Badgers spend the least time near the sett in the August–November period. Then they usually leave shortly after emergence and are away for most of the night. For example, Göransson found at a sett in Sweden that on 19 consecutive nights in September the badgers, after a short period near the sett, were away all night foraging (averaging about ten hours per night) before returning for another short period of activity by the sett before retiring. In Britain, the pattern is similar, but foraging is only continuous when food is scarce; when abundant the picture is very different. Cheeseman following badgers on wet nights using infra-red binoculars, found that after an hour or two's foraging on earthworms, they became satiated and would then curl up in the field, or more usually in an overground nest in a hedgerow, and go fast asleep for several hours. They would then have a further protracted forage before returning to the sett shortly before dawn.

So, as a generalisation one can say that feeding and breeding factors largely determine the activity pattern.

DIURNAL ACTIVITY

True daylight activity, as distinct from early emergence in the evening before the sun has set is not unusual, although it occurs in cubs rather more than adults.

In the June–August period, I have come across well-grown cubs foraging not far from their sett in the afternoon in deciduous woodland, where disturbance was minimal. Once, I was checking up on the best position for photography and focusing on the place where I thought the cubs would play, when I heard a familiar rustling and looking up, saw a cub returning. It disappeared down the entrance without a glance or smell in my direction, although I was only 3 or 4 m away. Cubs in secluded places undoubtedly forage occasionally during the day, perhaps to relieve the monotony of a very long spell below ground.

Adults, occasionally seen in the early morning, are usually making late returns to the sett after a night's foraging. Eileen Soper (1955) tells how she came upon a sow and well-grown cub returning as late as 10.00 on a brilliantly sunny August morning in Hertfordshire, but this made no difference to their time of emergence the following evening.

Sometimes badgers lie out away from their sett during the day in the shelter of bracken or brambles. If disturbed, they often make their way back to the main sett. This may have been the case when an adult was seen walking along the cliff on a sunny August day in mid-morning by all the passengers of a launch plying off the south coast of Cornwall!

J.F. Chapman also described how he watched an adult in mid-July on a very hot day 'travelling through scrub, snapping at insects and chewing grass as it went'. But the strangest case he witnessed was on a sunny morning in Upper Nidderdale, North Yorkshire, in winter following a snow blizzard. The badger was ambling through a wood towards a sett which had shown no activity over the previous eight days. Perhaps it was changing setts.

ACTIVITY PATTERNS

Shortage of food as in times of drought is commonly the cause of diurnal behaviour, but it is certainly unusual for a boar to break into a hen run at 14.00 on a May afternoon and kill six hens! Neville Baker, who described this incident, said that the killing attracted the attention of the gardener who followed the badger to its sett, from which it was later dug out.

Rob Tweddle told me of two interesting examples of feeding by day which occurred in Sutherland, Scotland. One was in early April at 14.30 when a badger was seen foraging on a hillside among boulders and the other was in September at 10.00 when an adult was watched for 20 minutes feeding on the carcass of a stag before running into a cairn on being disturbed. This incident occurred 650 m up in the mountains on a wet and windy day.

Most incidents of badgers seen in daylight are characterised by an apparent lack of awareness of the observer. This is well illustrated by an account given by R.R. Hershaw who noticed a badger at about 15.00 in bright sunshine loping alongside a hedge bordering a wood. He walked up to within 5 m of it to watch:

> To my surprise it seemed quite oblivious of me, for it took not the slightest notice, but nosed about rather like a scenting dog; in fact at one stage it picked up a stick in its mouth and ran backwards and forwards with it. I followed it for about 200 m before it was lost to view among some fallen masonry. It showed no signs of caution or awareness during the 20 minutes or so that I watched it.

Some daylight sightings may be due to the badger being poisoned or diseased. When dieldrin was used as a seed dressing there were several occasions when badgers were seen above ground in daylight behaving oddly. It was subsequently proved that they had eaten pigeons which had died from eating the treated corn (Jefferies 1968). Similarly during the investigation into the connection between badgers, cattle and bovine tuberculosis, several badgers were discovered wandering about above ground during the day. They were found to be badly infected (Muirhead *et al* 1974). However, it is not possible to be certain whether this diurnal activity was caused by the diseased condition or not.

When very small cubs are seen above ground in late March or April, the probable reason is that their mother has met her death and the cubs are desperate for food. Although not yet weaned, they venture out instinctively and go through the motions of foraging.

8 Food and feeding behaviour

Examination of a badger's skull (see Figure 2.9) reveals much about the diet of the animal. The incisors, canines and front premolars are typical of most carnivores, but the molars are much flattened and broad—a characteristic of many herbivores. In addition, the sagittal crest, the immense size of the temporalis muscles and the position and strength of the jaw articulation all emphasise the importance of the crushing and grinding process when feeding.

Investigations into the diet of the badger amply confirm that they are true omnivores, exploiting a great variety of both animal and plant foods.

A considerable amount of research has been done on the food of the badger, notably by Andersen in Denmark, Kruizinga in Holland, Popescu and Sin in Rumania, Notini and Skoog in Sweden and Likhachev and others in the USSR. In Britain, Barker has analysed some 4,000 dung specimens, Hancox over 2,000 and Bradbury about 4,000. I have also examined the contents of nearly 150 badger stomachs. In this chapter, I shall refer constantly to the findings of these research workers.

With all this evidence available, it is now possible to draw an accurate picture of a badger's diet sheet. In general, the results from the different countries tell the same story, although in detail they reflect the differences in availability of the various items.

It is easy enough to determine what a badger eats, but it is much more difficult to assess accurately the importance of each item in the feeding ecology of the animal. This can only be done by painstaking quantitative analysis of the food eaten and by correlating the results with the availability of the various items in the habitat.

Two main methods have been used to study the diet—stomach and dung analysis. Each has its advantages and disadvantages and a combination of the two produces the most comprehensive picture. More recently, direct observations of feeding using infra-red binoculars has added much to our knowledge.

Dung analysis is time consuming, but can be extremely interesting. It is not the messy process that the term suggests and can be recommended especially to those who have a flair for detective work. It is possible to build up quite quickly enough experience to make a useful assessment of a badger's diet sheet, although identification to species level needs specialist knowledge.

A useful procedure is as follows. First remove a small sample from the material for microscopic examination later. Then break up the remainder in a basin of water. If the dung is very dry it should be left to soak overnight. The material should then be placed in a sieve of 1–2 mm mesh (a gravy strainer is admirable!) and thoroughly washed under a running tap—preferably small amounts of material at a time. The cleaned remnants are then placed in a white dish of clean water and sorted out into food categories.

To help identification, a reference collection of hairs, feathers, skeletons, insects and seeds is invaluable.

The remains of plant material are often little affected by digestion as the badger is unable to deal effectively with cellulose, lignin or suberin. So husks of cereals, fragments of shell from acorns or nuts, seeds or skins of fleshy fruits, or pieces of bark can easily be recognised. Mammalian fur is not digested, but with large prey such as rabbits not much hair is swallowed as most of the skin is not eaten; but sufficient guard hairs are usually found to make identification possible. The recently born young of rabbits, rats and voles are eaten completely, but the hairs are so fine they can usually only be detected under a microscope. Fragments of bones and teeth also help identification, but with very young animals bones may be completely digested. Feathers and claws of birds pass through the gut unchanged and may often be recognised as belonging to a particular species. Insects, with their indigestible exoskeletons, leave more obvious clues such as the wings, legs and elytra of beetles and the bodies of bees, wasps or earwigs. When caterpillars and other larvae are eaten, their bodies are seldom fragmented so they too can often be identified accurately. Surprisingly, the egg-shells of birds are not usually dissolved by the stomach acids and fragments retain their characteristic markings and may be identified. The same applies to snail shells.

Soft-bodied animals, such as slugs and earthworms, are almost completely digested, but microscopic examination of the small sample taken from the dung before processing reveals the presence of earthworm chaetae and the radula teeth of slugs and snails.

For studying the frequency of certain food items in the diet, dung analysis is very useful indeed, but frequency taken alone can give a misleading picture. For example, a badger may pick up the odd beetle during a night's foraging at most seasons of the year, so in terms of its frequency in the dung, beetles appear to be an important food item. But, it is only over a comparatively short season that beetles are eaten in large numbers. To make a more accurate assessment, some food items can be counted and others measured by volume. The number of whole insects may be calculated from the quantity of legs, wings or elytra present in the dung, but with soft-bodied animals, such as earthworms, a different technique has to be used. Hancox calculated the approximate number eaten by counting under a microscope the chaetae present in a micro-sample of known volume taken from the main sample of dung, the volume of which was also measured. A similar technique has been employed for the radula teeth of slugs. The volume of cereal eaten can be calculated from the quantity of husks in the dung. To find the correct factor to apply to this calculation, it is possible to feed known quantities of cereal to captive badgers and subsequently find out how much appears in the droppings.

1. Sow with two well-grown cubs; July

2. Badgers playing; May

3. Boar, sow and cub relaxing after emergence; July

4. Adult erythristic sow

5. *Adult albino boar*

6. *Grooming session*

7. *In red sandstone districts, the coats of badgers soon become stained by the soil, through constant digging. The cub (foreground) has not yet acquired much colour*

8. *Sow bringing back a bundle of grass bedding*

9. *Badger in day nest, June*

10. *Boar setting scent on back of sow*

11. Woodland sett showing the ground outside worn hard and smooth by cub play. Mud marks may be seen on the scratching tree

12. Badger nest in a hollow tree with bedding overflowing; Sweden, August

13. An 'up-and-over'; made by badgers when crossing a hedgebank bordering a Somerset lane; March

14. Remains of a wasps' nest dug out the previous night; September

15. Sow with young cubs; May

16. Cub aged about eight weeks, at sett entrance; April

17. Sow suckling her three cubs

18. American badger, Taxidea taxus; North America

19. *Hog badger,* Arctonyx collaris; *South-East Asia*

20. *Honey badger,* Mellivora capensis; *north Kenya*

FOOD AND FEEDING BEHAVIOUR

GENERAL CONCLUSIONS

There is ample evidence that the badger is a true omnivore, taking both plant and animal foods in significant amounts at all seasons of the year. The range of food eaten is extremely wide, but reflects the characteristic food-finding behaviour of the animal. The badger is primarily a forager, not a hunter. Nose to ground, it searches persistently over long periods for any small prey it can find. Many of the items taken are found accidentally during this foraging. It may take many hours before sufficient food has been collected. Its aposematic colouring and noisy foraging gives early warning of its presence to the more active animals which can easily avoid capture.

Figure 8.1 Cubs foraging below a flowering elder

FOOD AND FEEDING BEHAVIOUR

Figure 8.2 Badgers on their way to their feeding grounds

The badger is not adapted for pouncing or fast running although it is quite capable of sudden darts at unsuspecting prey, so active mammals and adult birds seldom appear in the diet, but dead or unhealthy ones do. For example, adult rabbits suffering from myxomatosis are often killed.

With such keen senses of smell and hearing, badgers are able to locate the nests of young rabbits and rodents and detect the presence of insects in leaf litter or under the bark of decaying logs. When foraging, they repeatedly stop and listen between bouts of probing with their snout to discover any source of food by scent.

A source of plant food may be concentrated in a relatively small area, so that once discovered it can be collected in a leisurely manner. This applies particularly to cereal crops in the late summer and acorns in the autumn. However, in times of scarcity, especially during the winter, underground storage organs such as the corms of the wild arum (*A. maculatum*) are only found by persistent searching.

Although the badger's tastes are so catholic, it is obvious that it relies heavily on certain foods. Skoog groups the various items into a number of well defined categories. Under primary foods, he lists earthworms, insects, mammals and plant food, and under secondary foods, birds, reptiles, amphibians, molluscs and carrion. I am tempted to put carrion among the primary foods, because at certain seasons and in particular localities it is of great importance. However, I would agree that in places where alternative food is abundant carrion may only be of secondary importance. I would also break up 'plant food' into three categories, cereals, fruits and green materials, the first two being of primary importance.

FOOD AND FEEDING BEHAVIOUR

Figure 8.3 Frequency of occurrence (per cent) of main food categories throughout the year in 3,846 dung samples (from England, Scotland and Wales). By courtesy of Dr Keith Bradbury

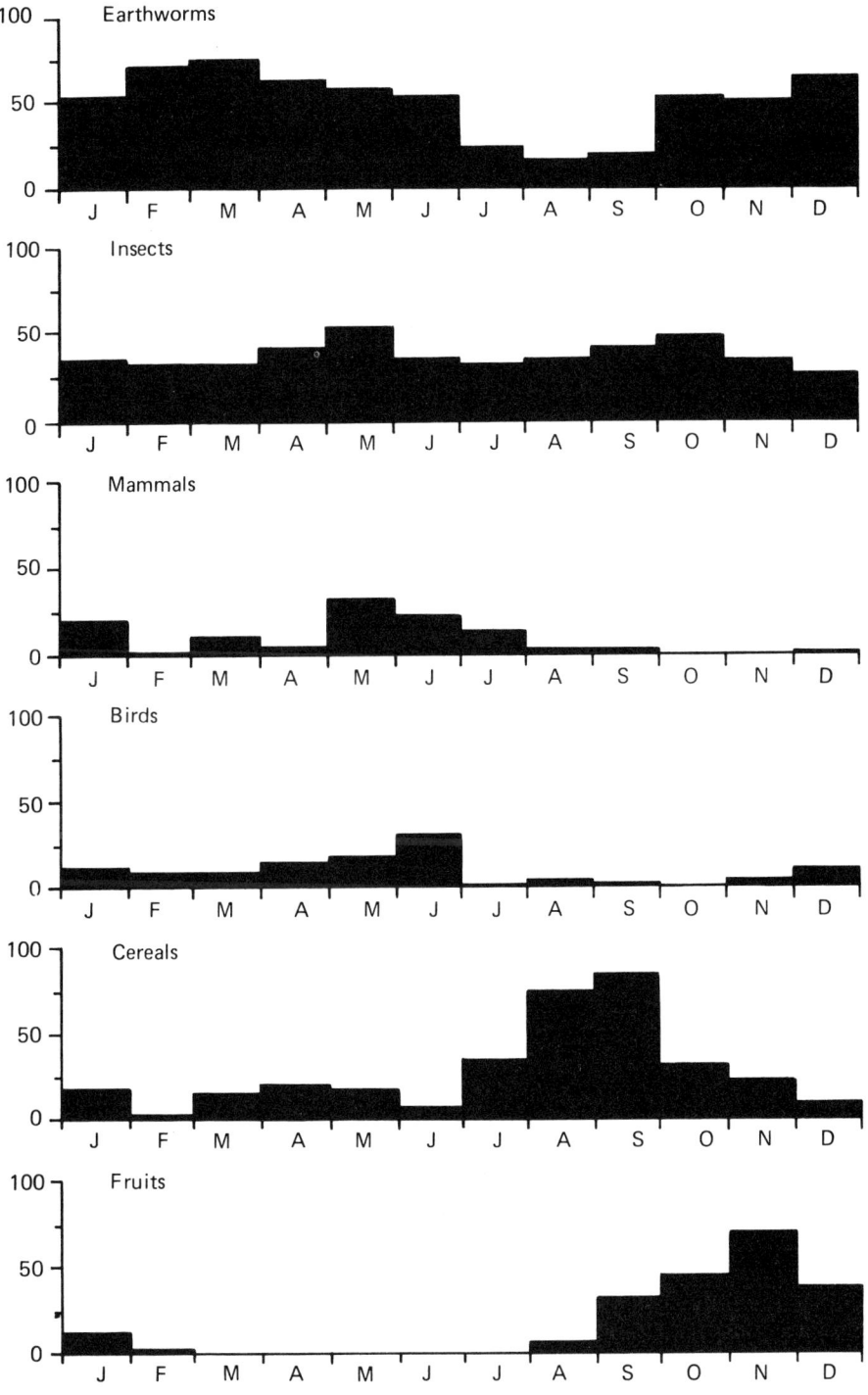

FOOD AND FEEDING BEHAVIOUR

EXPLOITATION OF HABITAT

The quantity of food obtained from each of the main categories varies enormously. When earthworms are available badgers primarily forage for these and for this reason Kruuk and Parish (1981) consider them as earthworm specialists. When earthworms are scarce or unobtainable badgers are opportunistic, exploiting whatever food is available in sufficient quantity to justify the search. For this reason their diet, although consistent for the main categories, differs markedly in detail according to the broad geographical location in which they are living, the type of local ecosystems that occur on their territories, the season of the year and, in the shorter term, the weather conditions prevailing at any particular time.

This is reflected in the way they exploit their habitat over the year. MAFF (1981) showed that badgers do not use the various habitats in their range in proportion to their size. Although permanent pasture comprised only 25 per cent of the area of each social group territory, in every month of the year badgers actively selected permanent pasture as a preferred foraging habitat, spending more than 50 per cent of their time foraging there. Arable land was visited in August and September when cereals ripened, but for the rest of the year was not a preferred habitat. Deciduous woodland was selected in spring and autumn whereas coniferous woodland was little used at any time of the year.

Variation with Geographical Location

In general terms, the further north the habitat, the longer and more severe the winters become. Hence, winter feeding in these regions becomes minimal and the badgers lie up in a state of semi-dormancy for long periods without feeding. This applies particularly to Scandinavia, northern Russia and Siberia, and in areas of higher altitude further south. By contrast, in milder and damper climates badgers feed actively during much of the winter and are less dependent on their stores of fat for survival.

For Britain, Bradbury maintains from his data that the diet in lowland areas becomes more variable as you go north because food finding is more difficult. I believe this is related to the number of nights when temperature and humidity are high enough for earthworms to appear on the surface of the ground. When these are available, other food items become of less importance.

Variation with Season and Weather

The effect of seasons and weather conditions cannot easily be separated as they are obviously inter-related. This is specially true in Britain, where some winters may be mild and damp and provide plenty of food and others may be severe and very little food of any kind is available apart from carrion.

As a broad generalisation, food of animal origin is of greater importance than plant food over most of the year, the exception being from mid-July to October when cereals and fruits of many kinds are of major significance.

Because earthworms are of such paramount importance in the feeding ecology of the badger, any factor that affects their availability such as mild damp conditions, with temperatures high enough to bring the earthworms to the surface, are of the greatest relevance. This is well illustrated by Barker's findings for Wytham Wood near Oxford over a period of two years

FOOD AND FEEDING BEHAVIOUR

(Figure 8.4). Here the volume of earthworms eaten during the wet summer of 1963 was many times greater than in the dry one of 1964 when there was a marked compensatory switch to cereals.

There is a distinct preference for earthworms over cereals whenever the former are easily available. During a dry spell in early August in Somerset the badgers were eating oats from a field near their sett night after night. Then came a short wet spell and they switched over at once to earthworms, although the oats were still available to them.

Under very severe drought conditions, especially in June, badgers may become desperate for food. This happened in many parts of England in 1975, 1976 and 1984, when earthworms were unobtainable and cereal crops were

Figure 8.4 Diet of badgers in lowland Britain in the wet summer of 1963. By courtesy of G. Barker (1969)

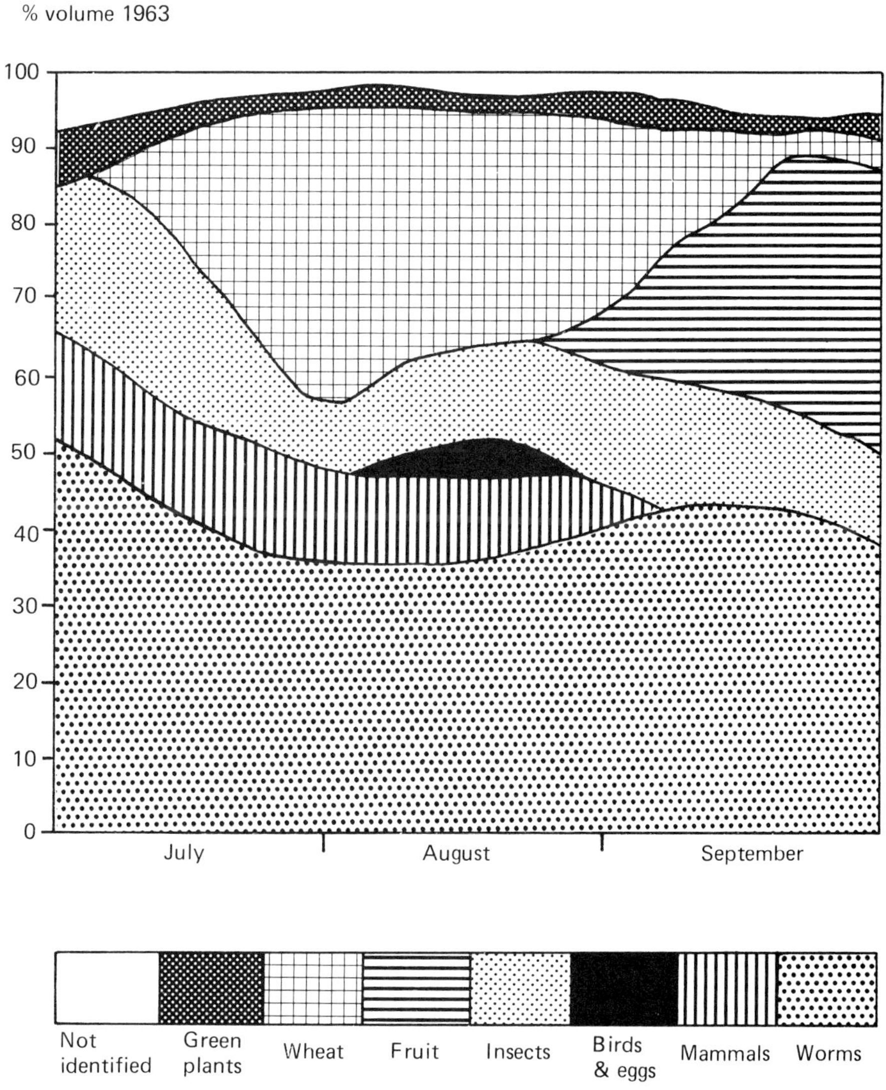

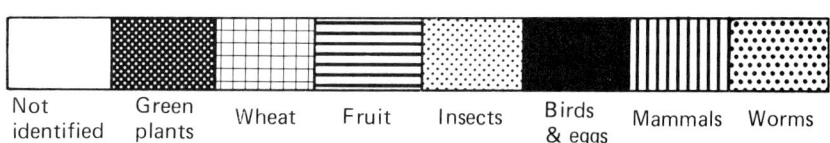

Figure 8.5 Diet of badgers in lowland Britain in the very dry summer of 1964, for comparison with Figure 8.4. By courtesy of G. Barker (1969)

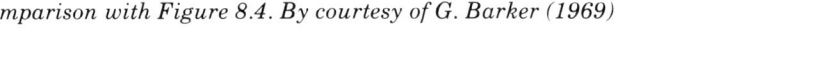

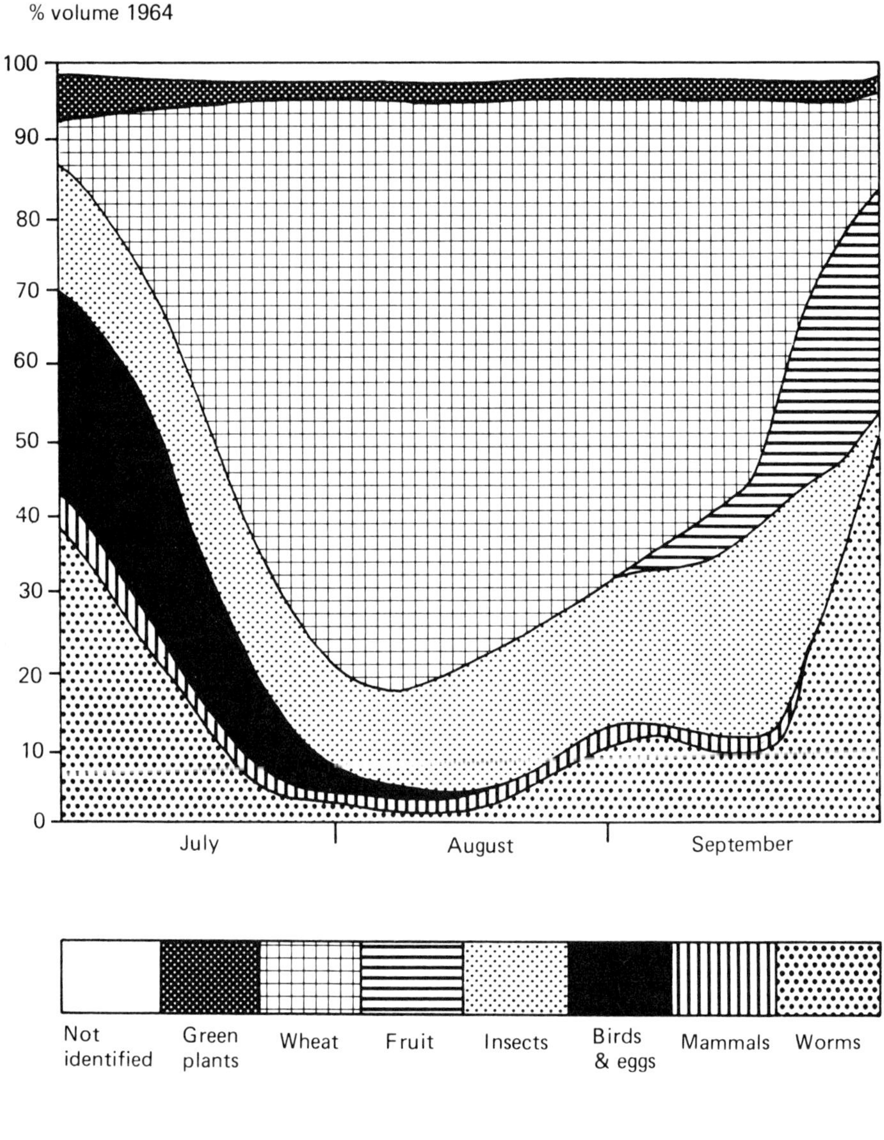

not sufficiently advanced to be exploited. Consequently there were many instances of unusual behaviour by badgers including raids on poultry, scavenging in gardens and much earlier emergences to lengthen the foraging period.

There is also a seasonal fluctuation in the badger's exploitation of mammals, birds and insects, again correlated with availability. More mammals and birds are taken in spring and early summer, when their young are numerous, and insects, such as beetles, become a more important food item in the late summer and autumn.

Variation with Type of Ecosystem

Badgers, being so adaptable, exploit many types of ecosystems, but they thrive best where there is a variety of habitats within their territories: this gives them a greater choice. This is one reason why setts near the edges of deciduous woods bordering arable and grassy meadows are so popular. When habitats are more uniform, population density tends to be lower, as in moorland situations.

The foraging area of a social group is often markedly circumscribed for territorial reasons, especially if numbers are high in the area (p. 148), so neighbouring groups may have very different sources of food available to them. Hancox analysed separately the dung from six separate home ranges in Wytham and showed clearly the link between species composition in the diet and availability, particularly for such items as wheat, acorns, blackberries and dor-beetles (*Geotrupes* sp.). He found the latter commonly in badger dung from home ranges which included pasture with plenty of cow dung, but they were virtually absent from those largely woodland.

With such variation among neighbouring groups, it is not surprising that the diet varies even more between badgers living in habitats of greater contrast such as those which are largely cereal producing and ones which are predominantly pasture. Again, badgers which have access to wetlands show considerable variation in their feeding habits. In Denmark, Andersen found that frogs and toads figured largely in the diet, especially during June and July, but in Britain they are relatively unimportant.

Moorland and mountainous habitats often deprive badgers of food items which are of major importance in more favourable places, but you usually find that there is a part of their territory where earthworms are abundant and large numbers of slugs, beetles and crane fly larvae (*Tipula* sp.) may compensate for the absence of other items. Carrion also becomes an extremely important source of food in such areas and fruits such as bilberries (*Vaccinium myrtillus*) are of great value in later summer.

Where food is relatively scarce and home ranges large, it is not unusual for badgers to move from one sett to another to exploit more fully a particular food which is seasonal in occurrence. They will move from woodland setts to those in hedgerows in August to be nearer supplies of earthworms and cereals, and badgers in mountainous districts may also descend to lower areas periodically where foraging is more productive.

THE MAIN FOOD CATEGORIES AND RELATED FEEDING BEHAVIOUR

Earthworms

The importance of earthworms in the diet of badgers in Britain cannot be overestimated. Without any doubt at all they are by far the most important single item of food. This has been amply demonstrated by the work of Barker, Hancox, Bradbury and, more recently, by Kruuk and Parish (1981). On the Continent the picture is the same. Only in extreme habitats where badgers are sparse, does the importance of earthworms decline to some extent. I agree very much with Barker that the availability of earthworms is the most important dietary factor in determining the population density in Britain.

FOOD AND FEEDING BEHAVIOUR

In Somerset, of the 115 badger stomachs I examined which had contents, 73 per cent contained large numbers of earthworms and of these, 62 per cent nothing but earthworms, apart from incidental material such as grass, which is always ingested in small quantities during the feeding process. In several stomachs, over 200 earthworms were counted and the volume of material exceeded 1,000 ml. In terms of weight, the highest was 642 g (about 23 oz) of earthworms in the stomach of a female in April.

In Yorkshire, Bradbury found 82 per cent of all dung samples contained earthworm chaetae and in at least 40 per cent they were the predominant food. He found earthworm remains in samples from setts in every situation visited, including those at high altitude, and on every type of soil.

Although earthworms are eaten whenever conditions are suitable, the period February–May is the most important. At this time, sows in particular need to feed well in order to suckle, and rapidly so as not to be away from young cubs for longer than necessary. On mild, damp nights, both these requirements are easily fulfilled as earthworms only lie on the surface when the ground is damp and when temperatures are above 2°C (optimum about 10°C—Satchell 1967). In many badger territories there are some patches which because of their situation are warmer on certain nights and these are selected for foraging as earthworms are more easily available (Kruuk *et al* 1979). Kruuk also showed that badgers prefer short grass pastures for foraging as earthworms are more easily captured there. Earthworm activity is at its greatest in the spring and late autumn, but the opportunities provided during the latter period are not exploited so much as in the spring because alternative food such as fleshy fruits and acorns are often available in quantity at that time. Nevertheless, in wet autumns earthworms are eaten in very great numbers, and also during the winter if the weather is mild enough. Relatively few are eaten in July and August. This is the period of least surface activity by earthworms and is also a time when other foods

Figure 8.6 The stomach contents of a single badger . . . over 200 earthworms

such as cereals, insects and early berries provide alternative food items.

Most of the earthworms eaten belong to the largest species, *Lumbricus terrestris*, which habitually lie out on the surface of the ground when feeding. Hancox found at Wytham Woods near Oxford, that the species comprised 75 per cent by volume of the earthworm biomass, although only 30 per cent of the total number of worms. As the biomass of earthworms in pasture can range from 1,000–4,000 kg per ha (Tischler 1965) there is an abundance of potential food available in good pasture if this ecosystem is included in their home range and the conditions are right for obtaining them. Species of *Allolobophora* are also taken, but in smaller numbers and occasionally other species are discovered under the bark of decaying logs and under cow pats.

Chris Ferris told me that in Kent some farms use irrigation sprays which run on cables. The sprays are used at night and very slowly travel the length of the field, rotating as they go. Both badgers and foxes make use of the wet conditions for worming when surrounding fields are dry, but whereas the badgers don't mind getting a soaking in the process, the foxes learn from the clicks made by the sprayer as it changes direction to avoid getting wet.

On wet nights, you can actually hear badgers feeding on earthworms if you approach near enough. A loud sucking noise reminiscent of spaghetti being eaten with relish, indicates success. However, when the worm is only partly out of its burrow it is gripped with incisors and tugged. If it breaks, the badger quickly drills after the remains with its snout, leaving a 'snuffle hole' to tell the tale. Worms are usually chopped once or twice by the teeth before swallowing, but sometimes are swallowed whole. Chris Cheeseman watched one badger in rich pasture eating worms at the rate of six or seven per minute without much pause for two hours, after which the badger curled up in the field and went to sleep!

Insects

Insects are an important source of food, being eaten as adults or larvae during every month. But only during periods of abundance are they eaten in quantity.

Only a few insect groups are of importance in the badgers' feeding ecology and these are almost all of large size and spend much time in or on the ground while feeding or breeding. The most important are the dung beetles (*Scarabaeidae*), wasps (*Vespidae*) and bumble bees (*Apidae*). Those of secondary importance include ground beetles (*Carabidae*), the caterpillars of moths, especially members of the *Noctuidae*, and the larvae of crane flies (*Tipulidae*).

Of the dung beetles, the large dor-beetles (*Geotrupes* sp.) are by far the most important. These insects are very active at night and visit the dung of cows and other herbivores. Badgers when foraging break up the cow pats with their claws or turn them over in order to find these beetles. Pits excavated beside or under the pats indicate where they have followed specimens which have burrowed into the ground. Over most of the year dor-beetles are only eaten in small numbers, but in May and August especially, large numbers may be consumed. Andersen in Denmark records over 800 in a single stomach in May!

The other dung beetles commonly eaten belong to the genus *Aphodius*. As their larvae feed near the surface of the ground in quite large aggregates they are often consumed in greater numbers than the adults (Skoog 1970).

Other beetle larvae which are favourites with badgers are the juicy cockchafer grubs (*Melolontha melolontha*) which feed on the roots of grasses and in some years may be very numerous in old pasture. Badgers may make regular visits to such areas and consume large quantities, night after night.

Sometimes these cockchafer grubs appear in lawns and golf courses, and if discovered by badgers, the shallow pits formed when digging for them may bring wrath on their heads. Under these circumstances, it is little comfort to the owner of the lawn to be told that the badgers are merely ridding the turf of insect pests! However, that is the case and the best way of preventing such damage is to tackle the primary cause, which is the presence of the cockchafer grubs.

Occasionally a smaller chafer (*Phyllopertha horticola*) may reach plague proportions, especially in upland pasture, and provide abundant food for badgers. C. Milner (1967) reported considerable damage to grassland over an area of 2,000 sq m in Snowdonia, North Wales, which he described as similar to that made by rooting pigs. It was not established for certain whether the badgers that caused the damage were after these chafer larvae or those of crane flies (*Tipula*), but the former appeared to be more probable. Wireworms are also eaten in quite large numbers in June and July in some areas.

The only other group of beetles found regularly in the diet is the *Carabidae*. These ground beetles are mainly nocturnal predators; a few are eaten during general foraging.

Some idea of the variety of beetle species eaten by badgers may be gained from those identified by Andersen in Denmark in the stomach of a single badger in June. It contained 452 individuals representing 35 genera!

Wasps' and bumble bees' nests become an important part of the badger's diet from July–September in some districts. They are probably located largely by scent, although hearing may play a part. As Skoog points out, wasps are an important food because they are common, are of a reasonable size, and are to be found in large numbers within a small area. A large wasps' nest is a complete meal for the taking. Badgers are extremely fond of wasp grubs. If one finds a nest it shows obvious signs of excitement, fluffing up its coat and darting here and there as it investigates the position of the nest. There is an old country saying that 'a badger never goes in where the wasps come out'. This is certainly true, as a badger will always dig down to

Figure 8.7 Beetle larvae are commonly dug up and eaten

FOOD AND FEEDING BEHAVIOUR

the nest from above, not expand the entrance used by the insects; this is just as well, as the one vulnerable spot where a badger may be stung is the end of its snout, the wasps being unable to reach the skin in other places when the hairs are erected. A badger will dig down very quickly with its powerful claws and make short work of adults, pupae and larvae as well as quantities of nest material. By morning, all that remains is a large hole in the bank, remnants of nest material scattered about and a few desultory wasps huddled pathetically in the ruins of their home. It is surprising how many adults are eaten during such a raid. Skoog records up to 300 in a single stomach. With these numbers it would seem almost inevitable that some stings would be received, although the quick chopping action of the teeth would kill the majority. I know of no evidence that badgers are immune to wasp venom.

Large numbers of wasps' nests are destroyed by badgers each season, which is one reason why badgers are popular among foresters, who are often plagued by them when clearing brush in the autumn.

The record for nest destruction probably goes to the badgers of Beaufort (Sussex) 'which devoured between them 30–40 nests during two ensuing nights' (Butterworth 1905). Badgers also occasionally pick up queen wasps during the early months of the year and in this way prevent the subsequent formation of nests.

Bumble bee nests are also destroyed in large numbers. These have the added attraction of honey to supplement the meal of larvae and pupae. Rather fewer adults are eaten than with wasps and they are chopped by the teeth more thoroughly before swallowing.

Sometimes hive bees may swarm from an apiary and perhaps make a nest in a hollow tree; occasionally such a nest is discovered by a badger. If the nest is accessible, it will tear the bark away with teeth and claws and in an incredibly short time devour much of the honeycomb. Very occasionally badgers may attack bee-hives when in isolated situations. Lawrie Webb told me of several instances where they had turned over hives to get at the honey destroying or reducing some colonies in the process. Arthur Dines (1981) also described how four strong colonies had been devastated by badgers in June. They appeared to have nosed the lids off the hives and abundant claw marks on the sides of the hives showed how they had separated the supers (boxes containing the honeycomb) from the broad chambers below.

Figure 8.8 The gape (shown here when eating a wasps' nest) shows how the nose projects beyond the lower jaw

The larvae of crane flies (*Tipula* sp.) may occur in sufficient density to reward persistent searching in the humus layer of woodland where the moss layer is turned over during the process (Skoog 1970). They may also be present in large numbers in damp pastures and upland grassy moors. They are eaten largely in late autumn, winter and spring, when they have attained sufficient size to make the search productive. The adults are also eaten at the time of a hatch. In August, Cheeseman saw badgers snapping them up from the pasture at the rate of 20 a minute!

Caterpillars of various moths, which feed at ground level, are eaten in fair numbers, although it is unusual for a badger to find many on a single night. The commonest taken is the large yellow underwing (*Triphaena pronuba*), but many others are eaten occasionally including those of swift moths which feed on the roots of grasses.

Ants are sometimes eaten when other food is scarce. I have seen ant-hills partly destroyed by badgers which had presumably been going for the larvae and pupae. Badger hairs left behind provided the evidence.

Woodlice, millipedes and centipedes are occasionally eaten but they are of little significance.

Mammals

Although the frequency of mammal material in dung and stomach contents is not so high as for some other items, there is no doubt that mammals constitute an important source of food. From the nutritional aspect a single mammal is the equivalent of a much larger number of smaller prey, and a nest of young rabbits can provide a substantial meal in a very short space of time.

The most important species taken are rodents (voles, mice and rats), insectivores (moles, shrews and hedgehogs) and lagomorphs (rabbits and hares). Badgers are not adapted to capture larger mammals, unless they are injured, old or diseased and even with the smaller species many more juveniles than adults are taken. Unexpected items occasionally recorded are squirrels, lambs, foxes and deer, but there are nearly always taken as carrion and will be discussed under that heading (see p. 131).

Mammals may be eaten during any month of the year, but the main peak is the spring and summer when most of the prey species breed. There may be a further peak in the winter, particularly if the weather is severe, but this almost certainly reflects the eating of carrion.

The commonest rodent taken is the short-tailed vole (*Microtus agrestis*). Even in situations where the bank vole (*Clethryonomis glareolus*) is common in the woods, *Microtus* is taken in much greater numbers probably because it is more easily caught. In fields, badgers can easily and effectively expose the runs of *Microtus*, which are often rather superficial, or dig down further to find the nest. A conical pit and a few badger hairs are left to tell the tale. In woods and hedgerows, where bank voles are more common, their nests are often in places (such as under tree roots) where digging is more difficult.

Badgers also occasionally catch water voles (*Arvicola* sp.) by digging out their runs. One excavation in Sweden was several metres long, 200 mm across and 300 mm deep. Badger hairs were found at one end where the vole had eventually been captured.

Wood mice (*Apodemus sylvaticus*) are eaten spasmodically. The adults are usually far too agile to catch, but when numbers are extremely high they are taken quite frequently (Skoog 1970). At other times only the nestlings are captured.

Rat nestlings also figure occasionally in the diet, but adults are seldom caught unless cornered.

Of the insectivores, moles are probably the most commonly eaten, both adults and young being taken. In the stomach of one badger killed by a car in the early morning, after a cold, dry night, I was surprised to find a lot of earthworms in addition to two moles. It is possible that the badger had found a mole larder in one of the runs, as no earthworms would have been on the surface of the ground that night. That badgers will do this was confirmed by Chris Ferris in 1983 when she watched a badger make for a cluster of mole hills. To quote from her account. 'It passed by some, deliberated at others, then coming to another, dug swiftly down and brought up a store of worms! Presumably the worms were discovered by smell.' Badgers will also excavate mole runs and may dig right into a mole fortress to get at the nestlings.

Shrews do not appear to be a favourite food of the badger. The common shrew (*Sorex aranius*) is the most usual species caught, but pigmy shrew (*Sorex minutus*) and water shrew (*Neomys fodiens*) have been recorded.

Hedgehogs (*Erinaceus europaeus*) are eaten occasionally. A.D. Middleton (1935) found four in the stomach of a single badger killed near Oxford, and remains have also been recorded by Bradbury and others. All accounts agree that remarkably few spines are swallowed.

There have been conflicting reports of how a badger succeeds in killing and eating a hedgehog without being damaged by the spines, but the most detailed eye-witness account comes from Chris Ferris. She was watching a female hedgehog with four young foraging among leaves when some badgers approached—two adults and three cubs. The hedgehogs all rolled up, the adult into a prickly ball, but the young making a poor job of it. The sow badger and her cubs promptly ate the young ones which presented no problem at all, while the boar took over the ball of spines. To quote from her description:

> When the hedgehog was on its back with the 'join' between head and hindquarters uppermost the boar made one swift movement of its left front paw and ran the claws straight down and into the 'join'. The hedgehog gave a sort of bounce and emitted a high-pitched cry that continued for some seconds. Then the right paw descended on the 'join' that was now open and raked sideways along the belly of the prey. The hedgehog was now opened flat and pinned at both ends by the formidable claws. The boar then lowered its head and began to eat.

On returning the next morning the skin was found to be flat, spines downwards, and apart from a tiny piece of head was quite clean.

When I have found the remains of hedgehogs, some skins had been turned inside out, others were flattened and a few looked like prickly balls with the spines outwards and no flesh remaining. R.W. Howard maintains that

> the skin is curled up in the normal manner of a frightened hedgehog and the spines are erected. The skin of the legs is sometimes turned inside out and sometimes absent; the jaws are invariably present, and the cranium sometimes. The skins form a complete ball and yet everything has been removed from the inside and no trace of blood or viscera is left lying about.

Chris Ferris suggests that the curling of the skin may be due to differential drying of the two surfaces.

Bradbury found very few hedgehog remains in his material. He makes the point that badgers and hedgehogs forage for rather similar prey and hunt in similar situations and are bound to encounter each other many times, yet relatively few hedgehogs appear to be eaten. He suggests that badgers either eat them reluctantly, or only certain individuals develop the necessary skill for dealing with them.

On the other hand, the skins of hedgehogs are not uncommonly found, and the question must be asked more critically whether the majority are killed by badgers or foxes, and if by the latter, how? Their claws are unsuitable.

Rabbits occur frequently in the diet, although not in such numbers as formerly. In pre-myxomatosis days in Britain nests of young rabbits were a staple part of a badger's diet during the spring. Normally, an adult cannot be caught by a badger, but in those days rabbit snaring was a common practice and badgers undoubtedly took their chances when they came across them in snares.

Since then rabbits have increased once more in many parts of Britain and Bradbury showed that nearly 7 per cent of his samples contained rabbit hair. In mild winters, the fur appeared in the dung even in January due to early breeding, but the peak was reached in April. Unweaned litters accounted for the vast majority.

Field evidence for this preference was given by Professor Tinbergen who said that each year a badger used to visit the Ravenglass sand dune region of Cumbria to feed on the rabbits. It dug up a number of rabbit stops (nests), but always seemed to choose the time when the young were near 'fledging'.

A badger locates a nest by scent or hearing—probably both. It digs down vertically in preference to opening up a tunnel.

Leverets are occasionally eaten, but they are far less important in the diet than young rabbits. Skoog describes how a sow and cub which were shot in Sweden were found to have eaten four young hares, three in the sow's stomach and one in the cub's. An interesting point was that the sow was carrying the leg of one of the leverets in her mouth when she was shot. Badgers typically eat their prey on the spot, as most prey is small, but under very unusual circumstances prey of relatively large size may be brought back (p. 83). One such instance was photographed by E.C.D. Darwall (Figure 8.9). He described the event as follows:

> On the evening of 17 June the boar went off into the woods surrounding the sett, leaving five cubs playing round the entrance. Ten minutes later it returned with a dead rabbit and I photographed it near the entrance. After the flash it took the rabbit into the sett. The rabbit appeared to be an adult, or at any rate a nearly fully-grown one. I think I heard sounds suggesting that the rabbit was being eaten, but I was distracted by the arrival of a fox which had obviously been following the scent of the rabbit, so that may only have been my imagination.

However, it is of course possible that it was the fox that first caught the rabbit and the badger somehow took it and brought it back to the sett, followed by the fox.

The strangest instance of food carrying was described by Janet Orchard (1958). On a farm in Oxfordshire, a new-born calf was missing. Investiga-

FOOD AND FEEDING BEHAVIOUR

Figure 8.9 It is extremely unusual for a badger to bring back food to the sett; this one is bringing a rabbit

tion by torchlight at a badger sett in a wood bordering the field revealed a cloven hoof sticking out from the main entrance of the sett. On pulling the leg the body of the calf was revealed. On returning with the calf to the cowshed, artificial respiration was given and the calf recovered. When it was daylight, Janet Orchard went back to the spot where the cow had calved and was able 'to follow the tracks of two badgers moving backwards as they dragged the calf across 200 m of frosty grass, under a barbed wire fence to which a few calf hairs still clung, and over 20 m of woodland to the main entrance of the sett'.

Suspected lamb killing by badgers has been reported from time to time, but very few instances have stood up to careful investigation. Usually other animals have been found to be the culprits. A Pest Control Officer told me of two instances he was asked to investigate in Somerset some years ago. On one he set traps for the killer and caught an Alsatian dog, and on the second, when going with the farmer to the orchard where the lambs had been killed, they caught the culprit in the act of repeating the crime—it was the farmer's own sow! Usually, the circumstantial evidence for suspecting badgers is that the remains of a lamb is found outside a sett, but when this happens it is practically certain that it is the work of a fox living there. Foxes habitually bring prey back to the sett when they have cubs, but badgers have no need to do so as they suckle their cubs for 3–4 months, by which time they are able to forage for themselves. However, still-born lambs are eaten occasionally.

Having said this, it does not follow that no badger has been a lamb killer. There are a very few instances—I know of about seven that have come my way over nearly 50 years—where the evidence is either certain or highly probable. In all cases, it would appear to have been the work of a single individual, usually an old boar.

FOOD AND FEEDING BEHAVIOUR

F.W. Baty (1952), Pest Control Officer for Gloucestershire, described how lambs were being taken apparently by badgers. One lamb was penned on the scene of the tragedies and an old boar badger was shot when about to attack this lamb. From the fact that there was no more trouble, it seemed that one 'rogue' was responsible.

S.P. Clarke told me of another instance on the Worcester–Hereford border where a farmer lost 21 lambs—usually one of each twin. Snow on one occasion showed up nothing but badger tracks. A badger was dug out and killed and there was no more trouble. Mortimer Batten (1923) recorded another instance in North Wales.

The method of lamb killing by badgers is distinctive. Badgers usually go for the hind quarters, especially the region of the back just in front of the root of the tail. This is also the position attacked when badgers are fighting between themselves during territorial disputes. Badgers are also said to crack the ribs, which a fox is claimed not to do. When eating the lamb, a badger goes for the guts and liver first and if a complete meal is made of the lamb, the skin is left inside out and often thrown over its head.

One can say with conviction that lamb killing is extremely rare and if it does occur is usually the work of a rogue animal which finds normal food difficult to get. It is not usually a hazard for a farmer and no action to destroy badgers on a farm is justified as a preventive precaution.

Extremely rarely, a rogue badger may attack a cat. I know of two instances where the deed was actually witnessed. However, normally cats are in no danger from badgers as the following incident, told me by E.N. Watts, shows. A stray cat which was quite wild lived at the bottom of his garden in Sussex and each evening at dusk it came to within 20 m of the house for a meal he put out for it. One evening, to his surprise he saw a badger sharing the meal with the cat, both eating from the same plate. They seemed very friendly and quite accustomed to each other!

Cannibalism occurs occasionally, but most of the evidence is circumstantial. I would speculate that the commonest cases are when a boar enters the part of the sett where there are very young cubs, and eats them, but certain evidence of this is lacking. Normally, the boar is banished from the breeding area by the sow and she is particularly careful not to go out until the boar has left, but accidents can happen. The late Lord Knutsford told me how he came across two very small cubs outside a sett which had just been killed and eaten out, except for their feet and heads. On digging out the sett, he found two old badgers, a boar and sow, which he was sure were the culprits. The sow was not in milk, so could not have been the mother. He came to the conclusion that the old pair had for some reason required the sett which had been occupied by a young sow and cubs.

There have also been other instances of dead cubs found outside setts with the skin completely cleaned out, except for some of the distal bones of the limbs.

Hancox came across a strange case of cannibalism which involved a senile sow and her three cubs, all of which were killed and partly eaten by a rogue badger. Nearby dung pits contained dung with cub hair and claws in it. The sow had been skinned inside out. He believes the killer may have been a half-tame sow which was released into the area previously.

Sows may occasionally eat their cubs if disturbed. This has certainly occurred in captivity. In general, the evidence suggests that cannibalism in badgers is exceptional and usually arises from abnormal circumstances.

Certain other mammalian species found in the dung or stomach contents will be mentioned under carrion.

ANIMAL FOOD OF SECONDARY IMPORTANCE
Birds and Eggs

Bradbury found wild bird remains in 10 per cent of his dung samples. At first sight this percentage seems surprisingly high as the badger is much more of a forager than a hunter. However, all workers agree that the great majority of bird remains found had almost certainly been eaten as carrion. The fact that gulls, crows, pigeons and even sand martins have been identified underlines this assumption.

Bird remains have been recorded for every month of the year although the majority occurred during the nesting period. However, Hancox showed that in his material pigeon remains went up dramatically in October, no doubt because of shooting. Badgers certainly pick up wounded birds and those which have subsequently died. There are also instances of badgers themselves dying after feeding on dead pigeons which had been killed through eating corn dressed with insecticide (Jefferies 1968). Rather more bird remains are found during the winter months, the species involved being mainly those which roost in woods and copses and which have died due to severe weather conditions.

During the breeding season, badgers undoubtedly come across the nests of ground-nesting birds, and eggs or chicks are taken. However, only in five instances out of 177 involving bird material, did Hancox find both an adult bird and a clutch of eggs in the same sample; and only in 16 was there evidence of fledglings. However, it is highly probable that young birds which have not yet acquired the strength for adequate flight and are sheltering on the ground may be predated.

It has been reported that badgers can sometimes rush at birds and catch them successfully. This is possible with ducks, especially during the moult, and sleeping waders, such as oystercatchers and curlews on the shore. However, Chris Ferris tells me that she has witnessed badgers kill and eat ground-roosting birds on bitter cold winter nights. The cold makes these birds torpid and easy to catch. Larks were the most usual in her area of Kent.

What about game birds? In Britain, the Game Conservancy considers that damage by badgers in pheasant rearing is insignificant and calls for no repressive measures against the badger. Pheasant nests are occasionally robbed by badgers, but these occurrences are exceptional and with normal badger densities this is no problem. Bradbury records pheasant egg fragments in only two dung samples. Hancox, working on an estate where pheasants are probably the commonest ground-nesting species, found the remains of 24 eggs in 2,000 samples. Adult pheasants have been recorded in material from Britain, Denmark and Sweden, but these are very rare events and in most cases the birds had probably been wounded by shooting, as some were cock birds, which do not incubate. However, some of the females may have been surprised on the nest, as Skoog mentions instances where remains of both adults and eggs occurred in the same material.

It has been the experience of many naturalists and game keepers in

Britain that pheasants have on numerous occasions raised broods successfully from nests very near occupied badger setts. Mrs Fairfax, writing in the *East Anglian*, said that during one season she knew of 14 hen pheasants which brooded safely within 100 m of a badger sett. Norah Burke also told me of a hen pheasant which hatched her full clutch safely within the perimeter of an occupied sett!

There is less data available regarding partridges, although I know of several instances where nests have been destroyed by badgers when foraging beside hedgerows. Damage of this kind does not appear to be a usual hazard.

Poultry killing by badgers has always been a subject of some controversy and emotion. It is easy to be prejudiced about the species as a whole if you have just lost some valuable hens or ducks to a marauding badger. It is also just as easy to convince yourself that badgers don't kill poultry if you happen to be fond of badgers and let your heart rule your head. Over the past 40 years I have had many letters about poultry killing from various parts of Britain and have attempted to sift the evidence and analyse the factors concerned. In addition, those who have laboriously investigated the diet of the badger through food analysis have contributed valuable factual data, so an objective judgement can be made.

The fact that badgers do kill poultry occasionally is indisputable. Apart from all the reported cases where foxes were clearly the culprits there are a number of eye-witness accounts, and many others where the circumstantial evidence is tantamount to proof. I will cite a typical instance in some detail, as the circumstances are significant. On 30 March in Hampshire, a farmer was awoken at 03.00 by a fiendish noise and rushed out into the snow to find a badger actually in the henhouse. It had presumably raised the drop hatch by getting its snout under it, but after entering, the door had fallen shut trapping it effectively. It had killed two hens. The farmer shot it and found it was an old boar with worn teeth. The next morning its tracks were followed in the snow. It had come from the local sett 300 m from the house, reached the garden and walked along the road. Its tracks then suddenly altered towards the henhouse some 30 m away and it was evident that it had bounded through the snow at high speed as if it had suddenly got wind of the birds. The farmer then told me that although he had kept poultry there for the previous seven years, he had never before lost any birds to a badger, although he had lost many birds to foxes.

I know of a number of other rather similar instances where a badger has obviously been attracted by the sounds and warm smells from a poultry house and entered by the same method of nosing up the hatch. On one occasion, on hearing squawks from the poultry, the farmer got out of bed and grabbed a gun and a torch. On opening the door of the henhouse, the badger rushed between his legs and escaped before he could fire a shot. To his surprise he found no hens had been killed. Incidentally, it is worth making the point here that if the drop hatch fits into a groove at the bottom of the door a badger cannot get its nose under to get inside!

I also know of several instances where badgers have forced their way into dilapidated henhouses by destroying the rotten wood, but usually when poultry are well housed and sensible precautions taken, little damage from badgers occurs. However, there have been a few instances of young poultry losing limbs when kept in 'arks' or houses which have a raised slatted floor. Sometimes a badger has been able to get underneath, get a grip of a bird's foot and pull, with distressing results.

Normally, when badgers kill poultry they do so in a very different manner from foxes. Foxes usually go for the neck, but badgers will attack the body. When badgers eat a carcass they often start at the vent, pull the guts out and eat that first; later they go for the more muscular parts such as the breast.

From all the evidence available, certain deductions may be made. First, poultry killing is an unusual occurrence. Bradbury found few instances of poultry having been killed although his dung samples came from all over Britain. The exception was from dung from a sett near Huddersfield from which traces were repeatedly recovered. He found that these badgers regularly visited a refuse heap near a poultry farm on which dead birds were thrown. A similar paucity of evidence is characteristic of the data from others in Britain and on the mainland of Europe. In many parts, poultry killing by badgers is almost unheard of, but in some parts, especially Ireland, South Wales and some countries with high badger populations, rather more instances occur.

Secondly, poultry killing is usually the work of particular individuals; it is not typical of badgers generally. In the majority of cases, if the badger is killed, there is no further damage although many other badgers may be in the neighbourhood. When killing does occur it is often the work of an old animal with very worn teeth, or one in bad health which is unable to feed normally. If such an animal discovers an easy source of food, such as poultry, it may acquire the habit and do quite a lot of damage.

Thirdly, poultry killing becomes more prevalent during times of food scarcity. A particularly important time is in February and March, if this coincides with severe weather, as the sows are then suckling cubs and their normal food items may be unattainable. An increase in poultry killing may also occur in times of drought. This was particularly obvious in parts of England during the severe summer droughts of 1975 and 1976 when many complaints were received, but they came mainly from regions of high badger density. As soon as the rain came and the earthworms were once more available the trouble abruptly stopped. This emphasises once more the spasmodic nature of poultry killing by badgers, when their normal food is available very little damage, if any, occurs.

To put poultry killing in perspective, Ralph Gibbons wrote to me about some badgers which had a sett under the floor of a garden shed in Watford. They had complete access to a hen run only a few feet away where there were four hens. There was never any damage to the hens and no evidence that any eggs were lost!

Badgers will occasionally take ducks and I know of one instance when a goose was killed, but again these cases are usually the work of an animal which for some reason finds normal food difficult to get. Phil Drabble (1969) described how he kept ten mallard nests under observation in a paddock where he had a semi-wild badger. They were all in places where the badger could easily find them. Every single nest hatched out and the adult birds got the ducklings to the pool.

However, not everyone is so fortunate. Miss M. Patterson, writing from Dyfed, described how a badger climbed over a low pigsty wall into one where 17 ducks were shut up. It killed and ate four of them and bit two others, one of which died later. It evidently had eaten so much that it was unable to climb back over the wall. It was found curled up asleep among its victims at 10.00 the next morning and was shot.

Mrs Lucas also wrote to me about an occasion when at dawn she saw a badger chasing her mallard round and round their pond until the pond was empty. The few ducklings that escaped 'were covered in mud and looked more like chocolate ducks'. She ended her account somewhat heroically by saying, 'Nevertheless, I prefer badgers to ducks!' Fortunately, duck killing is exceptional.

Reptiles, Amphibians and Fish

Adders, grass-snakes, lizards and slow-worms have all been recorded in the diet, but presumably badgers only come across them occasionally. E.D. Clements found a badger cub actually eating a slow-worm in daylight. At his approach the badger retreated, but Clements kept quite still and the badger soon re-emerged and for a full five minutes continued to eat it at his feet!

In Britain, there are very few records of live snakes being eaten by badgers, although tame badgers have eaten dead adders, with relish. But in parts of Sweden, remains of both adders and grass-snakes have occurred quite commonly in the dung (Skoog 1970). As far as I am aware, no encounter between a badger and an adder has been witnessed, but Ognev (1935) states that the venom has almost no effect on a badger.

Amphibians are eaten more frequently, especially frogs and toads; newts are taken more rarely. In Britain, amphibians are only of minor importance, but elsewhere in Europe they occur frequently in the diet. In Denmark, Andersen records that 14 per cent of the 190 stomachs examined contained frog or toad remains. Usually only one or two animals occurred in a single stomach, but on one occasion 17 toads were found. Frogs and toads were taken during each month from May to September, but were most commonly eaten in July. In two instances about 500 young frogs (just after metamorphosis) were found in a single stomach.

Frogs and toads are equally popular and are treated in the same manner before swallowing. E.M. Cawkwell described how he used to feed a tame badger on frogs. It would seize the frog, bite it hard and throw it out of its mouth as the frog urinated. It would then scrape its claws on it rapidly and crunch it whole in the mouth. Others have commented on this claw scraping, which is usually done extremely thoroughly. This would reduce the amount of obnoxious secretion which these amphibians (especially toads) produce from their skins.

Fish occasionally are eaten as carrion (see p. 132), but there is some evidence that they will catch them alive. Mark Fisher regularly took his tame badger out foraging after dark, and he told me how it found an eel in a ditch and ate it. He also watched it take minnows swimming in a bowl of water. Could badgers do this in the wild? The only eye-witness account that I know of was described by E.R. Brown, when a senior pupil at Felsted School, Surrey. It was recorded graphically in the Natural History Report for 1961 on which the following account is based.

After a prolonged badger watching expedition on 4 January, Brown decided to go home by a route which passed near the river, in the hope of seeing an otter. It was 01.05. As he approached the river he heard a fox bark and crept in that direction in order to catch a glimpse of it. He moved towards the river and suddenly saw a fish appear above the grass and a moment later heard it fall and flap about on the dry land. On approaching the river silently he was astonished to see a fully-grown badger standing on a solid mud spit which projected nearly 2 m into the water. It had its right fore paw raised and was staring intently into the water at its feet. Suddenly, it made a bear-like sweep with its paw and neatly flipped a small fish out of the water and onto the bank behind it. Brown watched fascinated for a few more minutes. Then the badger, apparently satisfied, turned and ate the fishes on the bank and wandered off.

Molluscs

Molluscs are frequently eaten in small numbers, but they do not constitute an important part of the diet. A possible exception to this is in moorland situations, where the large black slug (*Arion ater*) may be taken in greater numbers. Slugs are mainly eaten during wet weather and I have watched a badger on a wet night climb a tree presumably to find slugs. Before eating a large slug, a badger will roll it on the grass to remove the slimy exudation. Larger snails are treated similarly although the smaller ones are scrunched up at once.

Snails most often taken by badgers in Britain are *Cepaea nemoralis* and *Cepaea hortensis* (Hancox 1973). These are field snails of medium size. Garden snails (*Helix aspersa*) and Roman snails (*Helix pomatia*) are eaten less frequently.

In coastal regions, sea mussels (*Mytilus edulis*) may also be eaten. Skoog states that in the Baltic, the badgers evidently search the piles of seaweed for these bivalves. He described how one autumn, after a gale which had thrown up along the shore much seaweed with mussels attached, sea mussel remains occurred in large quantities in the dung.

Carrion and Refuse

Badgers will certainly take carrion. In Scotland, badgers have been known to feed on the carcasses of red deer, returning night after night until little is left. There is an account by MacNally (1970) of a dead hind calf in which poison had been inserted being placed on an island in a loch in the centre of a deer forest for the purpose of destroying the crows. The water round the island was deep and 20 m across, and yet a badger had scented the carrion and swum over. It was found dead after eating the poisoned meat. Badgers will also eviscerate calves if found dead and still-born lambs may be eaten occasionally.

Colin Russell used to feed his tame badger on grey squirrel carcasses and the badger's reactions to them was an interesting sidelight on how a badger would deal with a live mammal of that size. It first approached it with extreme caution, then a quick grasp by the teeth, a sudden sideways flick of the head and the squirrel was dashed heavily on to the floor. The squirrel was then skinned and by morning nothing remained except the complete skin.

Badgers visit starling roosts in winter to pick up birds which have died during the night, scavenge on dead rooks and pigeons after a shoot and visit refuse heaps on which dead poultry are thrown. They will also habitually search the tide-line round the coasts and consume dead gulls and any other animal matter they can find, including dead fish.

Skoog, writing of badgers living on islands in the Baltic, states that fish remains are often found in their dung as a result of shore foraging, and he used herring as bait very successfully when live trapping badgers in Sweden. This reminds me of the apocryphal story of how to catch badgers. A kipper is tied to a string and put down the sett; after a few moments it is pulled out again, with the badger on the end! Perhaps there is an element of truth in it!

Badgers are also known to have scavenged on salmon carrion. J.H. Cuthbert (1973) found that on the middle reaches of the Tweed during 1968–9 when salmon disease was prevalent, a number of carcasses were eaten by badgers. A large badger sett was situated about 200 m from the river and any salmon stranded within about 300 m up and down stream from the sett was rapidly disposed of. Frequently, the badgers dragged the carcasses into undergrowth well back from the river bank and consumed them there. He also found, on two occasions, skeletal remains of salmon near the sett.

It is difficult to assess the importance of carrion in a badger's diet, but I believe that a badger will usually take it if the opportunity arises, unless an abundance of preferred food is available, and that it is of particular importance in times of food scarcity especially during winter.

PLANT FOOD

Cereals

Cereals can be ranked as a primary food, but the quantity taken depends largely on availability, weather conditions and alternative food supplies.

In Britain, in good cereal-growing localities, husks in the dung usually start appearing in mid-July, and appreciable quantities are consumed until October while some gleaning may go on even after that.

Regions where less cereal is grown tend to have more pasture and a heavier rainfall, so the badgers eat less cereal and rely more on earthworms. For the same reasons less cereal is eaten during wet summers than in dry ones (see Figures 8.4 and 8.5).

Barker found that wheat appeared in dung samples much more often than any other cereal, but wheat is undoubtedly the most important grain crop in southern England from which most of the samples came. In Scandinavia oats is eaten far more than wheat (Andersen 1955), but there, oats is the main crop and wheat is only grown to any extent in the south. From my own observations in local situations in south-west England, where both oats and wheat have been available, I have not been able to detect a preference, although the badgers seem to start earlier on the oats, but this could be due to earlier ripening.

Barley is usually strictly avoided when other cereals are available; however, one badger's stomach I examined was full of it and in some districts the habit appears to be quite common. In 1967, at a sett near Taunton, there was a field of oats above the copse where the sett was situated, and a field of barley below. Badger tracks went into both, but in the barley they were merely pathways through it, very little was knocked down and there were no signs of its having been eaten; but in the oat field the tracks were more numerous and led to small areas where oats had been knocked down and grain eaten.

In another sett in a hedgerow bordering a barley field, badger tracks among the barley were clearly defined, but on analysing the dung from the resident badgers no barley was found, only earthworms. Presumably they had been foraging for these in the barley field.

FOOD AND FEEDING BEHAVIOUR

The technique of eating the grain varies according to the height of the stalks. In the drought of 1975 food was so scarce that the badgers went into the oatfields even before the grain had swollen. The stalks were only about 700 mm high. On 1 July, long before sunset, I was able to watch an adult boar feeding in the field. Sometimes it would rear up on its hind legs and snatch at the heads with its jaws, at other times it would bend the green stalks with its fore paws to bring the heads within reach.

Figure 8.10 Foraging in an oat field

Figure 8.11 When badgers knock down oats, the straws usually lie in a criss-cross manner

When the oats is taller and riper the badgers trample it down in patches and the straws are typically left in bundles which cross each other (Figure 8.11). The reason for this was told me by Kruizinga who watched badgers eating ripe oats in Holland. The badger would raise a front paw and take a sweep at some of the oats to pull the straws down, then pass the heads through its half-opened jaws, sieving off the grain. It would then take a sweep with its other paw and do the same, so the straws appeared crisscross. It would then move forward slightly, repeating the process, trampling the straws it had previously dealt with.

With wheat, the heads are usually bitten off if they are still green, but when ripe the grain may be extracted, leaving the husks still attached. In this connection Mrs Plummer, after noticing husks in badger droppings near a wheatfield in Kent, examined the crop carefully. Badger tracks went into it and some standing corn stalks were headless. At one place, where a badger path from the wheat entered the adjoining wood, she found a broken-off piece of elder tree around which was piled husks from the corn. Caught up in the cracks and crannies were husks and ears from the green corn. The amount of this residue increased for about a fortnight, by which time the corn was ripe and the badgers had ceased to use the stump. She concluded that the badgers somehow dragged the heads of corn through the cracks and by this means got the grain from the green husks. If badgers did use this technique it would be extremely interesting, as it would be an instance of tool using. I have examined photographs of the stump which are most convincing, but confirmation of how the grain was extracted is badly needed.

Damage done by badgers, through eating cereals, varies greatly according to locality. Only during the early stage when the crop is still standing is the damage significant, after that it is a matter of gleaning.

Skoog calculated that about 200 ml of oats was eaten per badger per night at the peak season in Sweden. In parts of Britain wheat or oats may be consumed in comparable amounts. However, as Bradbury points out only a small proportion of the total cereal eaten is from standing corn in a normal year.

Badgers are fond of maize and if this crop happens to be present on their home range, then some damage is likely to occur in the late autumn. Hancox showed that at Wytham, near Oxford, where maize growing was restricted to one area, residues only appeared in the dung of one social group of badgers. Kruizinga says that in Holland badgers select the earlier and sweeter varieties.

Badgers will even eat cattle cake if they can find it. W. Simpson told me how badgers, which regularly visited a farm in Gloucestershire, one night walked up the outside steps of a building where the cattle cake was stored in order to eat it.

Fruits, Seeds and Storage Organs

Much of the vegetable material eaten by badgers is in the form of fruits, seeds or underground food storage organs of the higher plants. These are of primary importance in the late summer and autumn. At this time, badgers are rapidly building up fat as a food reserve before winter makes feeding difficult. Much of this fat is derived from plant food, especially acorns.

Badgers are very fond of succulent fruit, especially if they are sweet. Badgers in some localities will switch their attention to blackberries as soon as they start to ripen, picking them off neatly with their mouths. Chris Cheeseman told me how this happened dramatically during the drought of 1976. He was watching them foraging using infra-red binoculars and noticed how the badgers picked off the ripe blackberries wherever they were in reach. Then one badger had a better idea and jumped off a bank right on top of a blackberry bush and was seen spreadeagled across it using a front paw to bend down the branches so that the fruit was within its reach and could be picked off one by one! No doubt with such a thick hide it didn't need to worry over the prickles!

In northern Europe they will take wild raspberries, and in hilly districts will concentrate on whortleberries. The latter are of primary importance in Sweden in the late summer (Skoog 1970). Elderberries are also very acceptable as soon as they drop to the ground and, if birds have been at the haws, the badgers will eat the fallen ones with relish, grinding up the hard stones. Yew berries are eaten greedily, the succulent outer parts being digested, while the poisonous seeds are either vomited or pass unchanged through the gut and so appear in the droppings. If badgers have access to orchards or gardens they will eat what fruit is available. Windfall apples, pears, plums, cherries and peaches have all been recorded, apples being particularly important in the late autumn.

Badgers have been known to climb wall-trained trees to reach the fruit, and being clumsy climbers they may cause damage by breaking off some of the smaller boughs in their efforts to reach the fruit. They may also visit vineyards to feed on the ripening grapes (Hainard 1961).

Badgers will eat wild strawberries and if the cultivated ones are available, they seem to consider this a real bonus. One lady living in Avon, near Bristol, was particularly unfortunate as the badgers not only ravaged her strawberries on a large scale, but also broke down the raspberry canes to reach the fruit, stripped the red currants and sampled the gooseberries. Only the blackcurrants and loganberries escaped! I hasten to add that her garden was near a wood where badgers were plentiful and that the damage occurred during the severe summer drought of 1975. Fortunately, incidents of this kind are rare, but I include it to show what badgers can do under exceptional circumstances. Preventive measures can be taken to avoid this kind of damage (p. 197). To put the eating of cultivated fruit in perspective, it should be stressed that under normal circumstances badgers seldom have access to soft fruit and in orchards their takings consist largely of windfalls.

Of the non-succulent fruits, acorns are the most important whenever available. Acorn crops are notoriously fickle, but in a good year they provide large quantities of food for badgers from late October. As acorns lie on the surface of the ground, they may be found as late as January and are even discovered by badgers under snow cover. One November, I watched a group of seven badgers foraging under the oak trees and picking up the acorns for hours on end. Each one was given individual treatment and thoroughly masticated before taking another. Beechmast and sweet chestnuts are less favoured, but are eaten in times of scarcity, and hazel nuts and walnuts have been recorded (Ognev 1935). Mavis Budd noticed how parts of the shells appear in the droppings when hazel nuts are eaten, so the kernels are not extracted before swallowing.

Underground storage organs of plants are eaten mainly during the winter. As early as late December, if mild, you can frequently see the pits where badgers have dug out and eaten the corms of the wild arum (*A. maculatum*). The shoots, which contain toxic oxalic acid, are bitten off and discarded. Badgers also dig out the globular tubers of pig-nut (*Conopodium majus*), and bluebell bulbs are eaten occasionally. Frost reported that badgers had eaten the bulbs of the related vernal squill (*Squilla vernalis*) from cliff turf in south Cornwall.

Onions, potatoes, carrots, beetroots, swedes and parsnips are eaten occasionally and of these those with a high sugar content are preferred. Sometimes badgers enter gardens and eat some of the bulbs.

I have found pieces of bark in a badger's stomach. In the very early months of the year, when the sap is rising, badgers occasionally strip off the bark of beech and sycamore trees near their sett and lick up the sweet exudations that result. The eating of bark is correlated with this behaviour. It is not a common practice and where it does occur very few trees are affected.

Green Food

The photosynthetic parts of plants are of minor importance and although leaves from a great number of species have been recorded in the dung, it would seem probable that most had been taken incidentally when foraging for earthworms. However, in winter and in very dry weather in summer, grass is sometimes eaten in large quantities, occasionally almost exclusively, but little is digested. During the drought of 1975, I watched adult badgers deliberately eating grass, which suggests that in summer at least it may be a means of taking in moisture. In severe weather in winter it might serve the same purpose.

Bradbury also found clover leaves in the dung, sometimes in the absence of earthworms and grass. On several occasions his samples contained virtually nothing but clover. There have also been reports of badgers eating kale and cabbage.

Fungi

The larger fungi are eaten occasionally but do not appear to be important items in the diet. Bradbury mentions traces of bracket fungi in winter and possibly of a species of *Morella*. Others have recorded mushrooms (*Agaricus* sp.), puffballs (*Lycoperdon* sp.) and various toadstools including species of *Boletus* and *Lactarius*.

DRINKING

In the wild, badgers drink from streams, rain puddles and cattle troughs. In regions where standing water is scarce due to the porous nature of the rock, badgers will drink from the rain water which collects in the hollows between branching tree trunks; mud marks on the bark often show where they have clambered up to reach it. In captivity, they drink regularly, lapping like a dog.

It does not seem to be essential for badger setts to be near a permanent supply of surface water, so it is probable that they derive most of their water requirements from their food.

MISCELLANEOUS ITEMS

Badgers do not confine themselves to the edible; they will chew at quite a wide variety of hard substances. Cubs will often gnaw at sticks at the sett entrance and badgers near golf courses even make a habit of chewing golf balls which they have retrieved from the rough! Even a cricket ball was

Figure 8.12 Badger drinking from a hollow in a sycamore

taken by a badger and left in its dung pit area covered in tooth marks. However, the most extraordinary case I have come across was told me by Croome Leach. He regularly watched a sett near Bristol and on one occasion he and his companion heard the badgers making a loud crunching noise near the sett—something like a dog cracking a big bone. It was too dark to see what it was eating. On shining the torch the badger was seen to eject something from its mouth before going back to the sett. On picking this up it was found to be a large jagged piece of bottle glass. There was no doubt at all that it had previously been scrunching up glass. Later, a piece of broken off top of an old-fashioned ginger-beer bottle was found down the entrance of the sett.

It would be nice to know why it was indulging in such a dangerous pastime!

SUMMARY

So, to summarise, I can only reiterate that the badger is a forager rather than a hunter. It is opportunistic, eating practically anything edible that comes its way, but it does have its preferences with earthworms right at the top of the list.

Figure 8.13 Drinking

9 Social life

Badgers are social animals which live together in groups in well-defined ranges. All members of a social group know each other as individuals. They may inhabit the same sett, or at particular seasons occupy neighbouring ones within the territory of the group. The territories of neighbouring social groups overlap very little.

This social organisation (which will be referred to in more detail later) is established by means of an elaborate signalling system involving visual, vocal and olfactory elements. Visual signals are only useful when the badgers are in close proximity, but vocal ones can be detected over much longer distances and are particularly helpful in darkness, or where undergrowth is thick. However, both visual and vocal signals do not persist beyond the very short periods when they are being used. By contrast, olfactory signals may last for many days or weeks and other badgers may react to them long after the one that made them has gone elsewhere. This lasting effect is most important in the social life of badgers and compares with the human activity of putting up notices when the occupier cannot be around all the time.

THE SIGNALLING SYSTEM

Signals are primarily for communication between members of the same species although some may be effective with other animals, and some olfactory signals may be of use to the badger that makes them, as when using its own scent marks to find its way back to more familiar country.

Visual Signs

The badger's poor eyesight probably precludes the wide use of visual signals, but they are used to a limited extent. In most carnivores, the animal's carriage reflects its mood and this is detected by others which take appropriate action. In badgers, one animal may stretch its neck forward and lower its head when another approaches. This may be a submissive gesture, as I once saw a sow do this when the dominant boar from a neighbouring sett came to the one she was occupying and the two came face to face. However, submissive gestures probably play relatively little part in the community life of badgers as all the members of a social group are usually friendly and

Figure 9.1 Frightened cub running back to sett, hair raised. Adults will also fluff up their coat as a threat and in defence

constantly make physical contact with one another. Hans Kruuk tells me that the lowering of the head may also be a signal that the badger is about to attack. The tail may also be used as an indicator as when it is raised vertically by a boar it is associated with sexual excitement. There is also the general fluffing up of the fur. This often occurs when cubs are excited, and as a defensive action when attacked. It may also be used by adults when attacked by dogs.

Vocal Signals

With such excellent hearing, it is not surprising that badgers make good use of vocal signals, although they can be silent over long periods. Vocalisation is much more typical of the period February–June as this includes both the main mating period when excitement among adults is at its height and the period when the cubs indulge in boisterous play. At other times vocalisation is less noticeable.

Badgers make a variety of threat sounds. There is a deep-throated, muffled growl which is a clear warning that an attack may follow. Growling also suggests annoyance, as I have heard them do this below ground when they are impatient to come out and know that I am there. This is not to be confused with the whickering of young cubs as they play below ground before emergence.

Another threat noise is the bark. This is a gruff, staccato note which is used if another badger comes too near when it is feeding: it may be preceded by a growl. The bark is often used when a badger is badly frightened, prior to headlong flight down the sett. I have heard it when a badger has come right up to me and suddenly realised my presence—first a sudden snort as it rapidly took in air through its nose, followed immediately by the bark as it ran away. Badgers will also spit like a cat when frightened. I once witnessed a prolonged fight between two badgers at Camberley in Surrey, during which growling, snarling and spitting noises were greatly in evidence.

Many people have commented on the badger's scream or yell. Unfortunately, errors of identification are easy to make, as the scream of a vixen is extremely variable and in some respects very similar to that of the badger. However, many instances described are undoubtedly authentic.

Frances Pitt vividly described it in this way: 'Badgers will make a most fiendish noise, uttering yell after yell of heartrending quality—scream after scream, long drawn and awful.' I can fully endorse this from personal experience; it is enough to make your hair stand on end!

The circumstances under which this screaming has been heard vary considerably and its meaning is still not fully understood. It may well be that with further recording and analysis the scream will be found to be a general term embracing several distinctive vocalisations, each with a different significance. However, the scream has great penetrating power so it can be assumed that its function is to carry some message over long distances.

On some occasions screaming is the result of severe fright. I have been told of wild cubs of about 15 weeks being picked up when they have wandered near an observer and each time this has caused the cub to scream piercingly and repeatedly until put down.

Kenneth Watkins used to let out his tame sow badger cub each night and on one occasion was awoken by persistent screaming. He found the cub (aged eight months) face to face with a large wild boar badger and in the morning the cub returned quite dazed and battered. This screaming appeared to be due to fear, as it took place before any attack was made.

Aubrey Seymour was awakened at midnight by a badger screaming just beyond his garden fence. The screams continued for some minutes, the badger only pausing to take short breaths. He described it as the sound of a beast in mortal agony. Five minutes later he heard another bout of screaming about 400 m away where there was a big sett, and while this was going on, he heard a badger pass by his garden in a great hurry, going in the direction of the screaming. It was in a great state of agitation judging by the low angry grunts it was making. It is possible that this screaming was a cry of pain or distress which caused the second badger to return hurriedly. This explanation is supported by another account of screaming being heard during the night and when investigations were made a dead badger was found lying beside the road in the direction from which the scream had come.

But the scream may be used as a challenge, as it has been reported most frequently between February and April, the period when territorial aggression by the boars is most evident. Screaming also occurs mainly where badger population density is high. But screaming is not confined to the early months but has been heard right through the year until November. My own view is that it is certainly caused by extreme fear, but also serves as a long-distance call, which may be a cry of pain or distress, or a territorial challenge.

A much more familiar sound is the whicker. This is characteristic of excitement and is most in evidence during cub play. It is a very variable high-pitched chatter, interspersed with stronger staccato notes—often when a cub is bitten rather harder than usual. Lower-pitched little growling noises may be incorporated among the more typical higher notes. Play between adults is often accompanied by rather similar whickering, but the sounds are not so high-pitched.

Very young cubs make a very querulous trilling chatter, which probably changes into the whicker when they become older. They make this when they greet the sow when she returns to suckle; it seems to signify an excited welcome.

Adults, especially sows, also make single high-pitched staccato sounds which have been likened to the note of a moorhen. I have seen cubs react to this call by quickly returning to the sow as she stood by the sett entrance. It also may serve as a contact call when badgers are travelling. John Whall has heard this particularly in January when the note was sometimes repeated two or three times. He has observed a badger making this call: 'it rears up on its hind legs, its front paws right off the ground and the head and neck stretched forward; it then drops on all fours, runs, and then repeats the process a few moments later.' This is very reminiscent of the smaller carnivores, such as stoats and weasels. When cubs are foraging over an area, I have also heard them make these staccato notes on finding some insect or other prey.

Finally, there is a series of vocalisations of varying intensity which I once likened inadequately to purring. The sound really has more of the qualities of a whinny, or a cross between that and a purr. I know of no single term to describe it, so will refer to it as a whinnying purr. A sow uses a very quiet form of this vocalisation when with young cubs and possibly when nursing them. It seems to express affection and at times reassurance, as when cubs are above ground for the first time. Captive animals will make the same noises when about to be fed and when they want attention. It becomes louder the greater the anticipation and excitement and may be made by both boars and sows.

Then there is a much deeper and more vibrant whinnying purr which is made by the boar and is very evident at the mating season. In February, when the sow is below, a boar may excitedly patrol the whole area making this vocalisation almost continuously. It also accompanies mating (see p. 174).

Scent Signals

'To stink like a badger' is a phrase not usually interpreted as a compliment, but to the badgers themselves scent is an essential part of life.

A badger's sense of smell is, to us, incredibly good. A badger's world is a world of smells and scent signals play a very significant part in their social life. In many instances their function is informative, in others they arouse emotions, but usually they do both.

There are various possible sources of scent which may be significant:

1. *The sub-caudal gland* (Figure 9.2). This is a large pouch, opening by a horizontal slit formed as an invagination of skin just under the tail. Its walls secrete a copious supply of a pale yellow fatty substance which is stored in the pouch. It has a rather faint musky scent. A badger sets scent or 'marks' with this gland.

2. *The anal glands*. These are two glands situated under the skin on either side of the tail region, which open by short ducts just internal to the anus. They secrete a darker yellowish-brown fluid with a very powerful rank musky odour, which is unpleasant when concentrated.

3. *The feces*. These have a very characteristic smell which is probably largely derived from anal gland secretion. So the feces may be looked upon as the vehicles for passing this scent to the outside world.

4. *Sweat and sebaceous glands*. These secrete a typical badger smell. It is often quite noticeable to the watcher just prior to emergence and may be the

Figure 9.2 Sub-caudal region of adult badgers

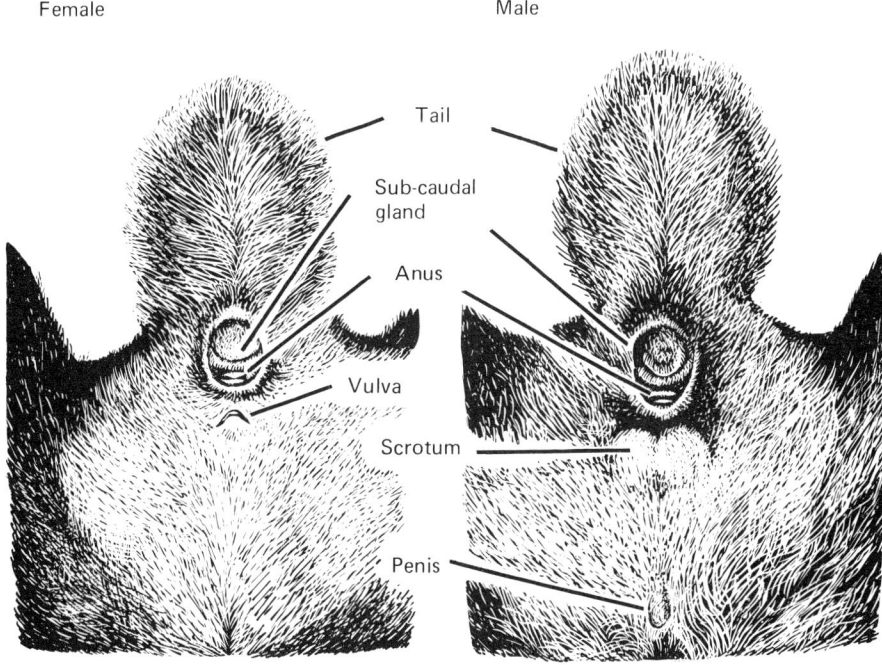

first indication that a badger is about to come out. This scent is particularly noticeable when badgers have been playing vigorously together and are hot—like dogs after exercise!

5. *Urine.* In some mammals this is the main scent marker, the leg-cocking action of the dog being a familiar example. There is very little information about its role in badgers, but the following incident is suggestive: Graham Madge was watching mutual grooming between several animals when the boar sprayed urine on more than one animal. This led to attempted mating. It is also probably important in establishing the oestrus condition of the sow, as some sex hormones are excreted in urine.

6. *Inter-digital glands.* These again are important in some mammals for leaving scent, but no work to my knowledge has been done to establish their presence in badgers. However, the claw-scraping behaviour on a tree near the main sett may be associated with the use of such glands and the scuffling action of the hind feet of a badger after defecation may be another example of their use. However, the feet may only be the means of distributing the scent.

Evidence is accumulating about the functions of the sub-caudal and anal glands and these will be discussed in more detail.

One of the characteristic actions by badgers is to set scent from the sub-caudal gland on each other. This 'marking' or 'musking' takes place between all members of the same social group, but most frequently by the dominant boar and particularly at the mating season. The tail is raised and the animal backs on to the other or even straddles it leaving a trace of the secretion on the fur. Martin Gorman *et al* (1984) showed by chemical analysis that this scent contains a number of fatty compounds (each with a characteristic smell) which are present in varying proportions in different individuals. Thus each badger has a personal scent. They also proved that trained animals could distinguish between scents of different badgers. So because musking takes place between all members of the group the latter will acquire a composite scent characteristic of that social group which is different from neighbouring groups. Hence a friend or stranger may be distinguished.

The personal scent may slowly change in composition with time. This is no problem to the members of the group as they repeatedly set scent on each other and so 'keep up' with any change. As Gorman pointed out this change could also be an advantage if a badger moved to another social group. Kruuk *et al* (1984) also suggest that the quantity of secretion produced is related to a badger's social status with dominant males producing most.

Musking of the sow by the boar is very evident during the mating season. It is also a means of familiarising a badger with its surroundings. H.R. Frank (1940) was one of the first to discover the implications of this action. Carrying his tame badger to country which was unfamiliar to it, he set it down and watched its behaviour, and very soon it was rooting about for food, oblivious of his presence. As it foraged, he noticed how it periodically squatted and set scent on various objects in its path, such as pieces of vegetation, stones or even the bare ground. These places Frank marked with small sticks. On taking the badger to the same area on subsequent occasions he found that it marked exactly the same objects with its musk. This repeated musking when travelling from a sett builds up a series of scent trails which act as highways for the badgers living there. They lead to places of particular importance such as feeding grounds. On reaching its destination the badger then leaves the main scent path and forages more widely. Later it searches around until it picks up the home path once more.

Some of these main paths become so impregnated with scent from generations of badgers that even if the field across which the path runs is ploughed up they will still be able to detect the scent and a new path will soon develop in the same place.

When badgers bring in bedding they do so backwards, and it is at first surprising that they unerringly find their way back to the sett without looking where they are going. The explanation is that they are keeping to a scent path. Similarly, a badger on its travels keeps its snout near the ground to pick up the scent more easily.

John Sankey told me that cubs brought up in captivity usually start to musk when about nine weeks old, young males showing the habit less strongly. This corresponds to about the time when cubs come above ground for the first time in the wild. A captive cub will musk on all sorts of objects, such as the shoes of its owner or an object in its pen such as a feeding trough. Living in an area permeated by its own smell seems to bring assurance and relaxation to the individual concerned.

When Frank's tame cub came across a place where a wild one had entered her territory through the fence, she at once set scent and did so on each of the snuffle holes the wild one had made in the course of foraging. This would appear to be an instance of making a personal mark which would inform other badgers that this was her territory. Similarly, a sow in early spring

Figure 9.3 Scent marking at dung pits

with small cubs below will often musk near the sett before going out to forage.

From January to May in particular, dung pits are established at strategic places over the home range, particularly at the perimeter and near main paths. These act as territorial markers, and anal gland secretion, when added to the dung, gives it a powerful scent; in addition, badgers often set scent from the sub-caudal glands nearby. So information is supplied by both glands. Badgers will react to foreign dung by defecating themselves as well as setting scent. In this, defecation seems to be an emotional response to the intrusion of another badger while the musking is like signing the protest. Boars visit these perimeter dung pits more than other members of the group.

Anal gland secretion is more volatile and direct emission of scent also occurs. This is a less subtle phenomenon and corresponds to the defence action of the skunk which squirts the fluid at the aggressor by violently contracting the muscular walls of the sacs. In the badger, this action is far less intense and only happens when an animal is suddenly frightened.

John Sankey mentions how some of his tame ones have done this and the smell has lasted on textiles for weeks after. I have also experienced this when a badger has suddenly discovered me and fled in fright. Incidentally, the teledu (*Mydaus javensis*) is called the stink badger because of the potency of its anal gland secretion.

COMPOSITION OF THE SOCIAL GROUP

The only reliable method of estimating the numbers present in a social group is to count the individual badgers. This is not difficult if the badgers are occupying a single sett and this is watched regularly throughout the year. But social groups often use several setts on the home range and the badgers often shift from one to another. This difficulty of counting was overcome in my early work (Neal 1948) at Rendcomb, Gloucestershire by posting watchers near each sett on certain nights and correlating observations. In this way, we were able to show over a period of three years that the social group occupying the four setts in Conigre Wood fluctuated between nine and eleven at the peak season when cubs were present above ground. However, another social group about 2 km away at Eycott used only one sett, but the numbers fluctuated more widely from year to year (7–12).

Figure 9.4 Badger family

During bovine tuberculosis investigations the Ministry of Agriculture completely removed a number of social groups in the Cotswolds, Avon and Cornwall. These gave precise figures of group size (Cheeseman *et al* 1981). Twenty-four groups were removed during the period June–October, mean group size including cubs being 5.6 (range 4.8–7.6), adults only, 4.2 (range 3.3–5.8); Kruuk and Parish (1982) working in four parts of Scotland estimated that for seven groups the average number of adults was 5.4 (range 2–11). So as a rule of thumb, five badgers per social group during the winter may on average be roughly correct for undisturbed situations. But it must be stressed that this is an average figure and fluctuations are wide, both from year to year and from one group to another. Some groups are consistently smaller than their neighbours and Kruuk and Parish (1982) suggests that this is related to the quality of worming areas on the range, modified by conditions which affect their availability.

The proportion of the sexes within the group varied considerably. Cheeseman found for his 24 groups the number of adult males ranged from 0–5 and adult females 0–7 with a combination of two males plus two females being the most common. Kruuk (1978) found that one group in Wytham Woods contained males only—two very old, three young adults and one middle-aged whose nose was missing, perhaps bitten off. This bachelor group may have been excluded from neighbouring ones. Cheeseman has not found bachelor groups in Gloucestershire, but he has recorded an all-female group which did not breed over two seasons and which he christened 'the nunnery'!

It is quite usual for more than one sow in a social group to have cubs in the same season. This has occurred even in high density areas where one might have expected some curb on reproduction. Cheeseman and Mallinson (1981)

Figure 9.5 Sow emerging

found in one group six cubs born to two sows in a territory of only 14 ha.

In semi-urban situations some setts become partially closed communities and this may give rise to abnormal numbers. Don Hunford in Essex had 22 in one sett in 1979 including at least three lactating sows and their seven cubs.

It is therefore possible in the spring to have in one main sett up to three breeding sows with their cubs, several mature boars and perhaps a few remaining yearlings. More often, however, when the group is large, it is broken up into smaller units which live in separate setts within the same territory, or if the sett is large and straggling, in different parts of the same one.

Sometimes the yearlings may be isolated in smaller setts, especially in the winter, while the main setts are occupied by breeding sows with one or two adult boars. The latter live in separate parts of the main sett when the cubs are small. If there is more than one sett within the territory which is suitable for breeding, and two adult sows are present, they will probably separate off, but if only one breeding sett is available they will keep together.

It seems likely that in an expanding population, social groups may gradually separate off as more setts are dug. This happened in a large sett in the Brendon Hills in Somerset, which I have kept under observation for 18 years. The sett is situated in a long, narrow copse 500 m long, which stretches between arable fields. When I first knew it in 1966, the badgers occupied the western end only. The copse is on a slope and the soil, formed from red sandstone, is ideal for digging. During the following ten years, badgers spread along the whole of the copse digging seven distinct setts. Two of these were seldom used for long, but foxes bred in one of them most years. Since 1971, badger activity has been concentrated around setts near each end of the copse where cubs were born most years. Towards the middle of the copse two setts have been used spasmodically, mainly by yearlings. Foraging patterns from the two end setts are now quite separate and the boundary between these two newly-formed social groups is well marked by dung pits.

ORGANISATION OF LIVING SPACE

How far do badgers travel? Wijngaarden and Peppel (1964) in Holland tracked them for as much as 6 km during a night, but the point of greatest distance from the sett was only 1.6 km. Rather similar distances have been recorded in Britain by Eunice Overend and Keith Neal.

It is clear from the mapping of badger setts in any area where the population density is high, that if outliers are ignored then the main breeding setts are spaced out at remarkably constant intervals. This is particularly noticeable in places where some geographical feature, such as an escarpment, causes the setts to be located in a linear manner. The higher the population density the nearer the main setts are to each other—a factor that is presumably influenced by the availability of suitable food in the area.

Kruuk (1978) found at Wytham Woods that the average distance between main setts was about 500 m. I have found similar spacing in the Brendons, Somerset, but here the entire region is so suitable for badgers that a saturation point for territories appears to have been reached. In small favourable localities much shorter distances between main setts have been recorded. In regions less suitable for badgers the setts tend to be spaced out according to the distribution of more favourable habitats.

In order to estimate range size, Skoog (1970) put down herrings as bait near main setts, incorporating tiny coloured glass beads. By examining the dung for traces of the beads he estimated the distance the badgers travelled.

Hans Kruuk at Wytham made further experiments. He placed food regularly near each main sett, incorporating coloured polythene pellets in his mixture of peanuts and syrup. A different colour was used in the bait for each sett. By examining the various dung pits for the coloured pellets he was able to map out the area covered by each social group. This enabled him to calculate that at Wytham, on average, each social group used an area of about 50 ha (Figure 9.6).

Figure 9.6 Territories of badgers of Wytham, as shown by recoveries in dung pits of coloured food markers presented on the sets (1–16). By courtesy of Hans Kruuk (1978)

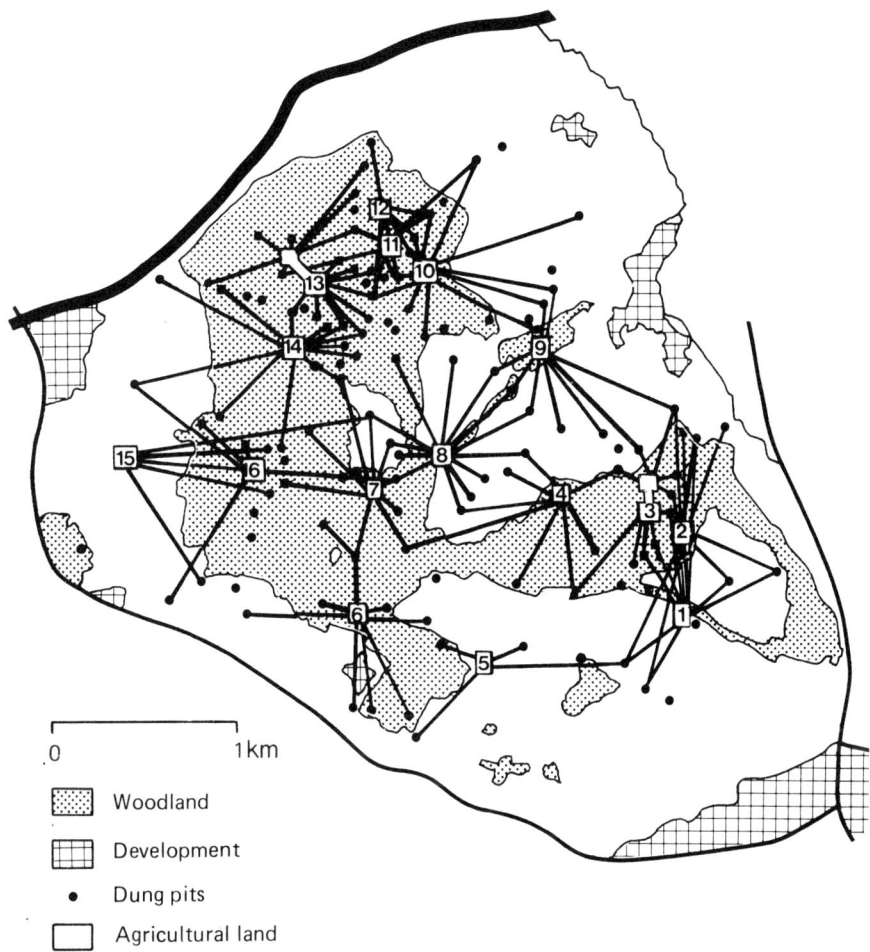

Since that time Chris Cheeseman (1979) working along similar lines in a valley in Gloucestershire, where badger density was unusually high, found that for the 32 territories each social group used on average, 40 ha, range 15–90 ha. This contrasts with regions of Avon and Cornwall where the average size was about 74 ha, and in four parts of Scotland (Kruuk and Parish 1982) 183 ha (range 120–309).

By combining radio tracking with bait-marking experiments Cheeseman showed that territories were not always exclusive as badgers sometimes trespassed. This was particularly evident at the breeding season when males made sorties into neighbouring territories, probably for mating. He also discovered that not all badgers in a group used the whole of the territory for foraging. Females in particular had smaller home ranges. Some individuals over a period of three months in the spring were using areas as small as 12 ha. Each individual had preferred areas.

Cheeseman also found that over several years of bait-marking experiments in Gloucestershire the territories varied little from year to year. Even when some groups were completely removed during disease-control operations the boundaries of the surrounding territories did not encroach on the cleared ones. A few badgers took up residence in the cleared area within a short period of time but their movements were more haphazard than normal. It seems that it will take several years before the population builds up and a stable pattern of territories is seen once more.

Kruuk found that badgers not only demarcate the perimeter of their ranges with dung pits, but the paths which often run between ranges are also impregnated with scent. Marking was often done on hummocks on which virtually every passing badger deposited a secretion of its sub-caudal gland. Badgers were also seen to rub their sub-caudal region 30–40 cm up a tree or post on the boundary while making a 'handstand'.

Areas delineated by territorial dung pits and perimeter paths may be looked upon as defended areas, at any rate during the first four or five months of the year. At this time aggression between badgers is not unusual and fighting can be very severe particularly in high density regions.

Kruuk described five fights, four of which were on range boundaries. There was a lot of biting, each ended with the intruder returning to its range. But more often badgers would avoid each other. Walking up to a boundary they either turned back or followed the perimeter path. Twice he saw badgers converging on a boundary and avoid contact by walking into their territories.

Don Bradnam cites one instance of territorial fighting between two boars when one died of its wounds days later.

Fights occur more commonly in regions of high density. Donald Hunford commented on this in Essex where urban spread has restricted the living space for badgers. Here a high proportion show signs of battle including torn ears and scars, particularly on the back above the tail. Many bite wounds have also been found on badgers in the Cotswolds, some infected with the micobacterium of bovine tuberculosis. So it must be concluded that the disease can be passed on in this way (Gallagher and Nelson 1979).

Although territorial fighting is more frequent between boars, it is certainly not confined to the one sex. When filming badgers at Camberley in Surrey I had purposely blocked up certain holes to increase the likelihood of badgers using the part we had illuminated for filming. I had not realised that in so doing I had forced another badger towards a part of the sett which was clearly the sow's territory. The intruder was seen to sniff at the entrance, but would not enter and as it turned, the resident sow shot out and attacked it. There was a tremendous fight as they twisted and turned and attempted to come to grips, growling, snarling and spitting as they fought. Eventually, the sow chased the other right through the wood. The noise went on for a full three minutes.

This and other rather similar evidence strongly suggests that for this short period of the year a sow with small cubs defends what has been described as a monopolised zone within a sett.

Aggression is much more marked during the first five months of the year and in particular the period February–April. Of the ten fights about which I have exact data, eight of these occurred during this period, the other two being in the autumn when there is often a renewal of territorial activity.

Also, there are many records of yearlings brought up in captivity being attacked by resident wild badgers. On some occasions they have actually been killed, but more often have returned severely battered, with nasty wounds. Again, a captive badger kept in a rural district will often attract the attention of the dominant boar of the local social group who will attempt to break into the pen in which it is kept.

So from these accounts it is clear that a resident boar will fight any intruder which hasn't the correct communal scent of his social group.

Some signs of hierarchy are evident within a social group, but its extent is uncertain. There is no doubt that in large groups there is a dominant boar—usually an older and larger animal. During the mating season in particular he may visit neighbouring groups and assert his dominance there too. It is probable that there is also a sow which is dominant to other females in the group who has first choice of breeding sites—again she is usually the older animal.

So to summarise, it appears that in high-density areas, each social group occupies a territory which is actively defended particularly during the first five months of the year. Notices to that effect are posted on the perimeter in the form of dung pits and scent marks and members of the group, particularly the boars, regularly visit these 'frontier posts' adding more dung, reinforcing the scent signals implying ownership. Later in the year territorial behaviour is far less evident, perimeter dung pits are used less and there is much more evidence of peaceful foraging. In summer and times of food shortage members of a group may go beyond the territorial limits without necessarily being molested. Within the defended territory lie one or more setts and if a sow has young cubs she may mark with her scent a small monopolised zone, often no more than a small part of the main sett. This she will defend against intruders.

Where conditions are less favourable this tight system of territories is less evident. Badgers have to roam more widely to find sufficient food, so ranges become very large. Boundary marking and patrol along the whole of the perimeter then becomes physically impossible and dung pits become restricted to such places as good worming patches or where a seasonal abundance of food is found. The pits then appear to serve as marks of possession.

Similarly in urban situations, Stephen Harris (1982), working in Bristol, found that territorial boundaries were poorly defined and home ranges overlapped much more.

One speculates whether territoriality is mainly a high-density phenomenon essential for conserving sufficient food throughout the year when competition is fierce, but when setts are few and far between this is less necessary and only vital food areas are marked. It should also be borne in mind that dung pits may also serve to provide other information concerning such questions as 'who is around' and his or her sexual state.

So the badgers' social system has many advantages. By having relatively large social groups containing more than one adult of each sex, an area can be defended more effectively. This is of great value during the breeding season and in times of food shortage. It is then that small ranges are so advantageous as the residents know the area so well that they can predict where food is likely to be found according to weather conditions. They can thus exploit a food source without unnecessary loss of foraging time and energy in random searching. If the sett is near the centre of the territory, access to all parts of the perimeter for defence and to different feeding areas is facilitated.

Figure 9.7 Dung pit area at territorial boundary

OTHER ASPECTS OF SOCIAL LIFE

Play

Badgers, in common with most carnivores, are very playful animals. Play is by no means confined to the cubs as adults frequently join in, and they also play on their own.

Cub play during the fortnight following their first emergence above ground is very tentative and mainly occurs just around the sett entrance. They keep in close contact most of the time, pushing with their snouts and toppling each other over more by accident than design. Sometimes the sow is with them and I have seen her lie across the sett entrance while her cubs scrambled over her body and played with her.

As co-ordination develops and they become stronger, they venture further

SOCIAL LIFE

from each other and from the sett entrance, but it is not until they are about 12 weeks old that play becomes really boisterous.

Play varies greatly in type and intensity according to circumstances. In secluded places where cubs are unsuspicious it may be vigorous, wide-ranging and uninhibited, but where disturbance is frequent it can be much reduced, especially in the vicinity of the sett.

Several varieties of cub play may be recognised. 'King-of-the-castle' is the usual type, when one cub will take up a position on the top of the heap outside the sett or on the bank above and the others will try to dislodge it and then take up a similar position. When bigger, the game may take place on a favourite tree trunk, but the idea of dislodgement is still there. Beatrice Gillam described how she watched three well-grown cubs in July playing on a horizontal tree trunk with a 2.5 m drop below. There was not a lot of room and falls were frequent, but the badgers appeared to come to no harm.

A variation is for one cub to emerge and then turn round and prevent the second from coming out. The first has the advantage of position and will playfully bite the one below each time it pops its head out. Eventually, the second comes out with a rush and there is a chase. Chasing is a characteristic form of play. It often occurs round and round a playing tree near the sett, so that after a week or so no vegetation remains. When one cub catches up with the leader, it will bite it on the rump or tail and then flee as the other rounds on it and takes its turn of chasing.

A further kind of play is when objects are used as playthings. Golf balls retrieved from the rough come into this category and sometimes old tins may be used. Occasionally, when two cubs play with the same object, a tug-of-war ensues as each fights for possession.

There are also times of particularly aggressive play. This often starts quietly but develops dramatically as excitement rises. One August, I witnessed this from such close quarters that two of the well-grown cubs I was watching collided headlong with my legs, nearly knocking me over. I felt I was a partner in the game as they played at my feet. One cub got hold of the ear of the other, then the tail and tried from below to tip it on to its back. They spun together in tight circles, trying to bite the other's tail and then both rushed up the bank. More scrapping took place there and one lost its balance and went head over heels down the slope like a furry ball. Quickly recovering, it climbed the bank once more and immediately they were at it

Figure 9.8 Rolling position

again. Occasionally, one would leap-frog over the other and if knocked over would roll on its back and bite upwards at its opponent. All the time there was a constant whickering of excitement, with intermittent louder yelps.

Sometimes, cubs will grip each other in a mutual bite, when the lower jaw of one is gripped by the jaws of the other. As Eric Ashby remarked, a twist by one badger will put the other on its back as a result of this grip.

Incidentally, rolling down the slope like a ball was a popular pastime of Mark Fisher's tame badger. He writes, 'I have been amused to notice that when rolling down a steep slope she can stop however fast she is going to save herself from crashing into a wall or beck.' He also told me how a wild badger was seen to escape from dogs in the fells by rolling down a scree. Stories of badgers rolling on gin traps were also quite frequent and some appeared to be authentic.

Analysis of such boisterous play suggests that typically the main components are:

(1) Chasing.
(2) Attempts to bite and hold the side of the neck, ears and tail.
(3) Attempts to turn the other upside down by using a low approach and then giving an upwards or sideways movement of the head.
(4) Rolling over on to the back in defence.
(5) Flight.

Sometimes during bouts of play cubs will momentarily show displacement activity and 'attack' vegetation. J.F. Chapman watched them pulling up bracken and shaking young trees in their excitement. He also described how there was a large sloping stone outside the sett's main entrance which the cubs used to slide down over and over again.

Play between adults has many of the ingredients of cub play. Early one May, I watched an adult boar and sow playing together quite aggressively near the sett. They started to play head on and mouth to mouth with their mouths half-open, twisting and turning in mock fighting without closing their jaws. This led to aggressive attempts to get a grip on ears and neck and as the excitement grew they turned in tight circles as they went for each others' rump and tail. It seemed significant that whenever play stopped temporarily the sow always had her back to the sett entrance. I found out a week later that she had cubs below.

Play clearly has a number of different functions. It obviously promotes physical development and co-ordination and many of its features train the animal for more serious fighting later without causing harm in the process of learning. It may also help to get rid of surplus energy. However, badgers are very social animals and I believe an important function of their playing together is to strengthen the bond between members of the same social group by constant physical contact. Play between adults probably also has sexual significance, although mating may often take place without preliminary play.

Grooming

It is a familiar sight to see badgers grooming after emergence. This they do with great thoroughness, contorting themselves in various ways to get at every part of their anatomy. Typical positions are shown in Figure 9.9.

SOCIAL LIFE

One of the more amusing postures often shown by an adult boar is when he sits fairly upright on his haunches, possibly leaning slightly backwards and scratches his belly with slow deliberate actions of both front paws. He will sometimes do this while sitting in the shallow bowl-like depression often found near a sett entrance.

Katharine Tottenham told me a rather intriguing story of a badger she brought up in captivity. It was given straw for bedding, but on one occasion this contained a dried thistle. It sorted it out, appeared to comb it upwards from the root with its front claws and then rolling on its back proceeded to scrub its chest with the prickles, holding the plant between its front paws. This appeared to be a deliberate action, because following this incident Mrs Tottenham regularly used to cut large green thistles for it and as soon as they were put into the pen it went through the same drill exactly!

Figure 9.9 Grooming attitudes (drawn from life)

Although scratching may be a natural response to parasites, and possibly to midges, it also appears to bring satisfaction to the scratcher. Sometimes it persists for half-an-hour or more! Mutual grooming also takes place and has a social function. The badgers nibble at each other's coats, and when several are in a tight huddle it doesn't seem to matter who does it to which. A sow will also carefully groom her cubs and may hold one down with her front paws while she goes over its fur methodically.

Phil Drabble (1971) described how mutual grooming regularly took place between an adult boar and sow in his artificial sett before emergence. His tape recordings showed that quite often the grooming lasted for at least half-an-hour and was accompanied by low whinnying purrs of affection. He suggested that mutual grooming might help to distribute scent from one animal to another.

John Sankey told me that if the skin on a tame badger's back is rubbed with the fingers, or vigorously combed, the animal at once starts to groom itself, a piece of cloth or a human hand, whichever happens to be easiest to reach. The teeth are run over the surface and hair is drawn through them, but there is no attempt to bite.

Badgers will also lick their fur like a cat, although they will not use a paw in order to clean their face. Licking often takes place when they come back from foraging. Mutual licking also occurs occasionally. David Mitchell wrote how two adult badgers (boar and sow) faced each other and then proceeded to lick each other's face, neck and back.

What happens when Badgers Die?

It is strongly believed by some countrymen that badgers bury their dead. Good evidence is hard to come by, but several instances of badger funerals have been published. Unfortunately, the most detailed and dramatic accounts have been related at second hand (Hampton 1947; Vesey-Fitzgerald 1942) and it is difficult to know to what extent they are accurate. The common factors in both these accounts were that a hole was dug, the body was dragged to it by more than one badger and earth was heaped on top.

It is certainly established that badgers will drag the body of a dead one for some distance. Bronwen Doncaster related how she saw a badger laboriously drag another across a road one February night. In spite of interruptions by passing cars, the badger persisted and eventually dragged the body up the far bank, where it was found the next morning. There is also an account by Joseph O'Kelly (1969) of a badger, killed by a car, being covered in leaves on two successive nights, presumably by badgers. The leaves had to be dragged some distance from a copse in order to do this.

I once found a dead cub aged about ten weeks outside a badger sett. I did not touch it, but watched that night to see what the other badgers would do with it when they emerged. In the event, it was an anticlimax; several adults came out and passed within 1 m of the body but they took no notice of it and went off!

All that can be said at present about badger funerals is that if they do occur, they are very rare events. But badgers are remarkable creatures and it is well to keep an open mind about the possibility. I'm still hoping that one day I shall see a badger funeral myself.

One thing is quite certain. The majority of badgers which die naturally, die in their setts. On several occasions diggers have found the remains of dead badgers in side tunnels which had been walled in with earth, presumably by other badgers.

No doubt when this happens and the body has decomposed, others will re-excavate that part of the sett and the bones will be pushed out with the soil—perhaps years later. This is the logical explanation of why old badger skulls are often found on the spoil heaps outside well-established setts.

10 Reproduction and development

Although the breeding cycles of mammals differ considerably in detail, they follow a broadly similar pattern. The succession of events may be summarised as follows. At some period when the individuals are mature, mating takes place. If fertilisation follows, the egg undergoes cell division, to form a hollow sphere of cells known as a blastocyst. This early development takes place as it passes down a narrow convoluted tube which leads from the ovary to the uterus. In most species, the blastocyst then implants in the wall of the uterus, which has been made receptive by the action of hormones. However, in some species the blastocyst may remain in the uterus for a period ranging from a few weeks to eleven months before implantation takes place: this phenomenon is known as delayed implantation. After implantation, a placenta is formed and the fetus develops as usual, the period of active gestation between implantation and birth being remarkably constant for each species. In all mammals, after birth, the young are fed exclusively on milk from the mother, this period of lactation lasting from a few weeks in small rodents to several years in the elephant. The time taken to reach maturity also differs greatly according to species.

Variations in this general pattern of events usually reflects the adaptations of the species to such factors as the seasonal changes of its habitat, the availability of food, its size and body weight, its ability to protect its young, and so on. Reproduction in the badger is of particular interest in these respects.

For many years the details of the breeding cycle of the badger were the subject of much speculation and controversy, largely because observations on a nocturnal animal were difficult and partly because the cycle was by no means typical. It was not until field observations were correlated with anatomical and physiological investigations that some of the complexities were unravelled. I would like to record with much gratitude the outstanding help received from Professor R.J. Harrison, FRS in this connection.

For the sake of clarity I will first deal with the birth of the cubs, as other events in the breeding cycle are to some extent governed by this happening.

BIRTH

It is not possible to be certain of the exact dates when badgers are born in the wild, because birth normally takes place in an underground chamber.

REPRODUCTION AND DEVELOPMENT

However, there are now a number of records of birth dates in captivity and a significant number of accounts of litters discovered as a result of digging and of cubs of only a few days old, from which times of birth can be estimated. A less reliable means of estimation is to take the first time when cubs are seen above ground and work backwards, assuming that they are seldom seen before eight weeks old. This method provides useful corroboration but is only accurate to within about two weeks. Another way is to judge the dates from the size of embryos found during post-mortems on badgers.

Figure 10.1 Three-week-old fetus within the uterus

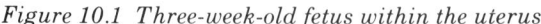

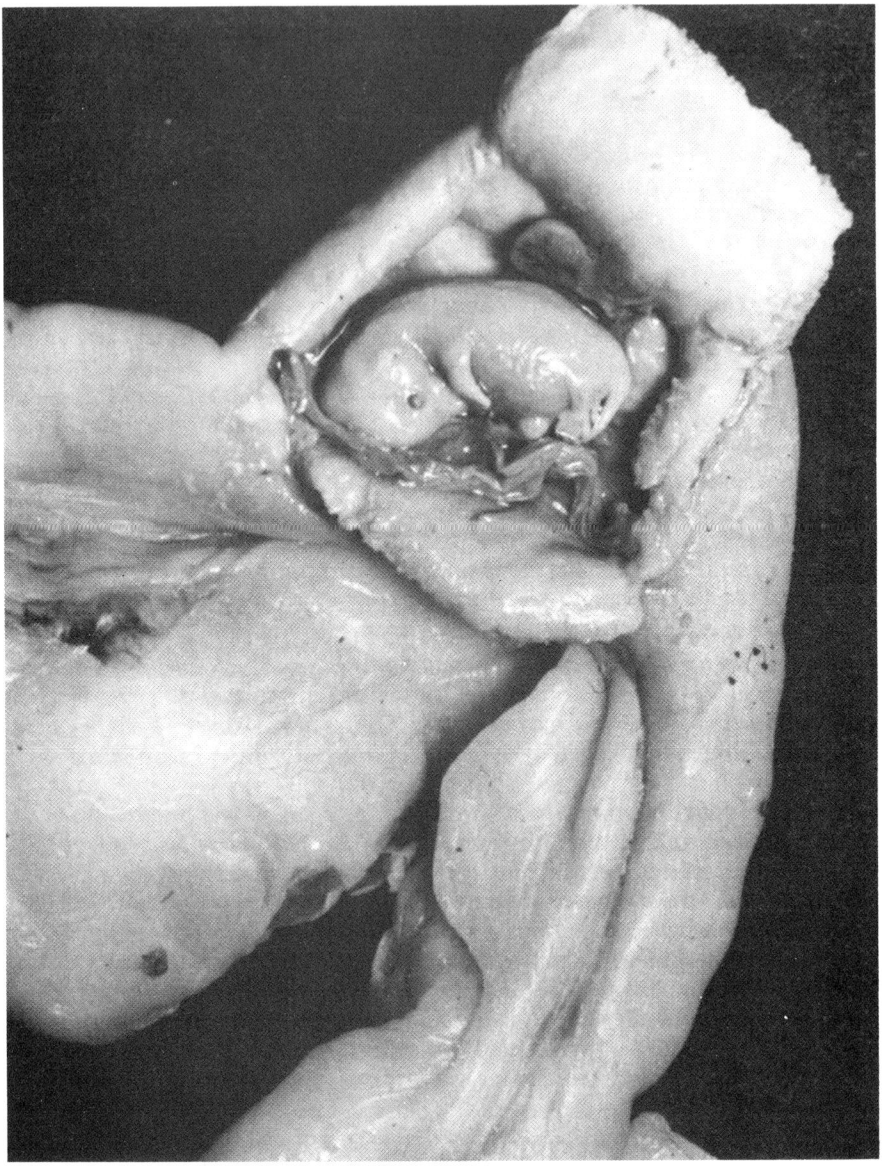

Figure 10.2 Birth dates of badgers from south-west England (in half months) and estimated start of weaning times at three months of age. Weaning time often corresponds with a second mating period

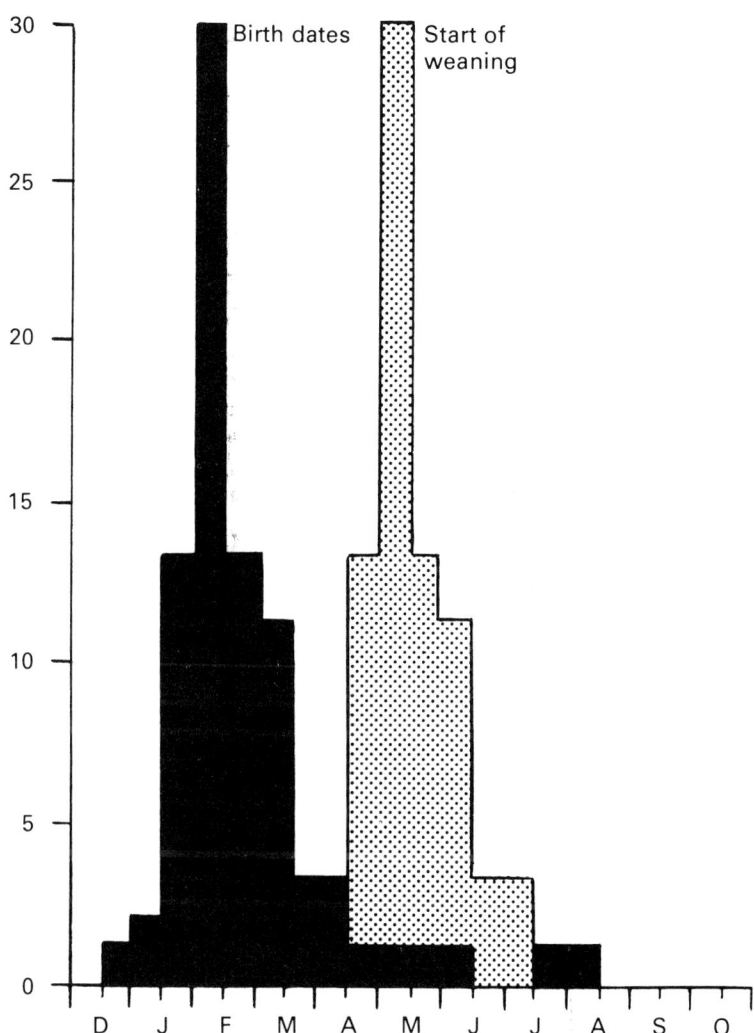

Taking data from all these sources for 88 litters in southern and south-west England, the great majority of births (76 per cent) occurred between mid-January and mid-March (Figure 10.2). The peak period is the first three weeks of February, but late January and early March dates are quite frequent and December births certainly occur. Neville Barker told me of a terrier which brought out a small cub on 1 January 1960 near Reading, Berkshire. He estimated the cub was a few days old and there are accounts of cubs being seen above ground on 18 February in Devon by N.R. Soultanian and on 25 February by David Humphries near Cheltenham in Gloucestershire. Humphries adds that the two cubs had obviously been up for some time as they were quite confident and chased each other down a bank. Both these instances point certainly to December births. However, any births in

December and after April can be looked upon as exceptional. The latest in the year records are from Devon, where two families, each of three cubs, were dug out on 29 July and 5 August 1956, their age being estimated by Mrs R. Murray at 3–4 weeks. A cub with eyes still shut was also recorded from Germany in June by Frank (1940).

There is a correlation of average times of birth with latitude. For south-west France the average date calculated from embryo size in Canivenc's material is 31 January; for south-west England it is 8 February: for Yorkshire, towards the end of February (Paget and Middleton 1974b) and in Scotland the peak period is probably early March; in Germany (Frank 1940) and Sweden (Notini 1948) is early March and in Russia (Ognev 1935) it is late March, with April for the Caucasus. There is also evidence which suggests that the time of birth also varies more locally according to altitude.

The number of cubs in a litter varies from one to five. To find the average number born, good evidence is obtained by counting the fetuses found in dead sows and the placental scars left from the previous pregnancy which are easily seen as dark patches on the inner surface of the uterine wall. These persist for several months after parturition. Direct evidence is also available from litters born in captivity and from those occasionally discovered in the wild. From a combination of these data for south and south-west England I found that from 37 litters, 8 per cent were singles, 16 per cent twins, 54 per cent triplets, 7 per cent quads and 3 per cent quins, with an average of 2.92 per litter. The average litter size for Europe as a whole is 2.7. Both litter size and pregnancy rate are lower in yearlings than older animals.

Many badger watchers have recorded the number of cubs per litter seen above ground some 8–10 weeks after birth. For 110 litters, 11 per cent were singles, 52 per cent twins, 29 per cent triplets, 6 per cent quads and 2 per cent quins with an average of 2.36. Compared with my earlier figure of 2.92 this represents a 19 per cent loss during the period below ground. It is of interest that at birth the greatest number of litters are triplets, but at 8–10 weeks twins are the most usual. It would seem from these figures that the loss of one cub from a litter of triplets is quite usual. Wandeler and Graf (1982) have shown that in Switzerland a number of sows in spring showed fresh placental scars, but no signs of suckling offspring, suggesting that a number of whole litters may not survive.

The nest chamber is an enlargement of one of the tunnels, often a side one and it may have a second tunnel leading from it. It is filled with bedding preparatory to the birth of the cubs.

The breeding chamber is often situated not far from an entrance. This has the advantage of better ventilation and might also enable the sow to isolate a small part of the sett system from the rest of the occupants and have an entrance to herself for coming and going. When filming badgers at Camberley, condensed water vapour was seen rising like steam from one entrance on a frosty morning in February, due to heat from the badgers' bodies; bedding used for the cubs was taken down the same hole, and later the cubs were heard being suckled approximately 4–5 m from that entrance. This is no isolated case, as it is often possible to predict the entrance the cubs will use when they eventually come above ground by noting the one into which bedding is taken during February and March.

Cubs up to three or four weeks old when brought up in captivity need much warmth if they are to survive, so you would expect the temperature of the nest chamber to be comparable. In February a maximum/minimum thermometer fixed to a long springy wire was pushed down a sett entrance where cubs were thought to be present. It is unlikely that the wire reached the breeding chamber, but a rise of 8°C was recorded, compared with outside. On the same day in an unoccupied part of the sett there was only a rise of 3°C at a similar distance down the tunnel. It seems probable to me that the temperature of the inside of the nest may be as high as 18–20°C.

For the first few weeks the cubs are buried in a mass of dry bedding which is an excellent insulator, so the heat from their bodies is retained. This is likely to be increased at intervals when the sow returns to suckle them. However, she does not usually remain with them during their first few weeks, but lives in a separate part of the sett during the day. This may explain why dogs sometimes enter setts and bring out very small cubs without being molested by the mother.

Very occasionally, breeding is known to occur above ground in Britain. The first that came to my notice was in a hawthorn hedgerow on a peat moor in Somerset (Neal 1969). The farmer, Sam Musgrave, heard whickering noises from a large heap of grass and reeds and wondered whether it was an otter's nest. When he poked it with a stick, a sow badger burst through the roof and stood grunting at him before slowly retreating. Inside he found several small cubs. I saw the nest a few days later, by which time the sow had taken the cubs away; the roof had gone, but it was still 500 mm high and more than 700 mm in diameter. The grasses and reeds seemed to be roughly woven together, giving strength to the walls. It was understandable that the sow could not dig into the peat, as the water-table was nearly level with the surface.

On another occasion (Neal 1969), a Somerset farmer, Jack Richards, discovered a nest in a large lean-to shed butting on to his barn. It was used for storing rough timber planking. It was dark inside and as we stood there listening I was astonished to hear the unmistakable whickering of small cubs coming from the far side. On climbing over the timber, the noises stopped, so I switched on a torch and peered down between the planks. There was a sow badger lying at full length on top of a great heap of hay. The light made her eyes sparkle but she did not move; she was on her side and was suckling a cub! After some time she slowly stood up, shook herself and without undue hurry made her way between the planks into the open barn next door. As we watched, the cub moved round and round, burying itself in the heap of hay. I moved the top covering gently with a stick and found two cubs curled up together. They were about 4–5 weeks old. On being exposed they started to pull the hay round them and were soon buried again.

In both these cases of breeding above ground, the sows appeared to be young animals and it is likely that they had been unable to establish territories in the more typical badger country not far away.

THE NEW-BORN CUB AND EARLY DEVELOPMENT

The average length of newly-born cubs is about 120 mm from the tip of the snout to the root of the tail. The tail is usually another 30–40 mm.

Weights are much more variable, ranging from 75–132 g in Britain. Those from large litters are usually lighter. G. Dangerfield found in a litter of four that the males weighed 90 g and 80 g and the females were both 75 g, while twin fetuses, estimated at full term and weighed by A. Killingley were 132 g for the male and 130 g for the female. Other weights for southern England were 84 g and 100 g for males and 89 g for a female.

A newly-born cub has a pink skin which is covered with greyish-white silky fur, more sparse on the ventral side. Darker hairs can usually be made out in the position of the eye stripe and on the lower part of the limbs. I have seen signs of the facial stripe in full-term fetuses; however, in some cubs stripes only become visible some days after birth.

At birth, the eyelids are fused, a state which lasts several weeks. Dangerfield has bred many badgers in captivity and has found that the eyes open consistently around five weeks. It does not follow that because the eyes are open the cub can use its eyes, as at this stage they are normally living in complete darkness underground and will continue to do so for several more weeks. Observations on cubs brought up in captivity suggest that under light conditions they are unable to focus on near objects for some weeks after their eyes are open. However, by the time they would normally be above ground they appear to notice objects within a radius of a metre or so (p. 90).

When 6–7 weeks old, cubs are ready to start exploring the underground tunnels, but seldom come above ground before eight weeks, and are often not seen until 9–10 weeks old. The badger watcher is fortunate to see a first emergence as it usually occurs some time after the normal time for adults to leave the sett.

Sows will occasionally move small cubs from one sett to another. Graham Madge (1982) described such an instance. In early April he was watching a large sett of about 27 entrances. Most of the holes were on an open hill site, but three were on the flat nearer the river. A sow emerged from one hole about 30 minutes before dusk and disappeared down another, 10 m from it. She soon re-emerged carrying a cub in her jaws by the scruff of the neck and hanging quite limp. She carried it at speed down the slope and almost threw it into one of the lower holes. In a minute or so she was out again and repeated the performance with three cubs in succession. The whole removal took 14 minutes and was carried out in an urgent and purposeful manner. Madge estimated that the cubs were 4–6 weeks old. Why the cubs were transferred is a matter for conjecture, but another litter of cubs was seen later at the same sett and the removal may have been to prevent interference by the other sow which could have been dominant to her.

Continuous all-night vigils for ten consecutive nights between 18–28 April 1953, gave a good picture of early behaviour. At that time Professor Hewer and I were filming badgers under artificial light and we took it in turns to man the camera throughout the night. To quote from our account (1954):

> Cubs were first seen on 25th April, although previously their shufflings and whickerings had been heard down the tunnel. Two cubs first appeared at an entrance at 22.50, keeping in close contact with each other. They were constantly on the move, investigating each stick or lump within reach and testing out their sense of balance on the sloping sides of the entrance. They were visible for about 10 minutes on the first occasion. They appeared again at 01.35 and were out for longer. The next evening they were seen at 20.40 for a short time and again at 01.00 for 25 minutes. On each of these occasions the sow was

below ground, but near the entrance, and at times her low whinnying purrs could be heard.

When only one cub is present, the first emergence is somewhat different as the sow brings up the cub herself. On 28 April at a sett on the Quantock Hills, I saw a sow lying just inside the sett entrance. A cub appeared from below and nosed about close to her body but would not venture further. However, when she got up and walked a few paces away it followed her and then crept under her body where it remained half hidden as she slowly walked away. Not long after, it left its mother and scampered back to the sett.

D.B. McGregor described a rather similar occasion when he saw a tiny cub poking out from below its mother at a sett entrance on the Mendips. This time she kept nosing it down the hole, apparently to keep it in.

The date when cubs first appear above ground is related to the time of birth. April is the usual month in south and south-western Britain, but much earlier appearances have been recorded (p. 161).

When 9-10 weeks old, cub appearances above ground become much more regular. They may emerge before it is dark, their activities become more energetic, and as their powers of co-ordination develop their play becomes more variable and purposeful.

At this age, they learn very quickly, many associations being built up, especially in relation to scent, touch and hearing. Early on, any sudden noise results in a scamper for home, but they soon learn to discriminate between sounds and no longer react to familiar noises.

Cubs become more venturesome when 10-12 weeks old and constant assurance is of less importance. They will explore the near environs of the sett and play becomes boisterous and prolonged. This causes the ground outside the sett to be beaten flat and appear almost polished.

A number of adult activities are foreshadowed during development. Setting scent on objects near the sett may be seen occasionally, although it is difficult to observe in the wild owing to the low position of the body. Tame cubs do this constantly. They also start making snuffle holes with snout and claws in their search for food. By July, trial holes are dug out which may not go in more than a metre before being abandoned.

Shuffling backwards with leaves and sticks is often observed from nine weeks onwards. At first, it is only momentary, but later the action becomes sustained and develops into the habit of bringing back bedding to the sett. E. Nettleton found that if cubs were separated from adult badgers when about ten days old, they nevertheless carried out all these actions in a similar way. This suggests that setting scent, forming snuffle holes, digging and bringing back bedding are inherited patterns of behaviour and not the result of learning.

GROWTH RATE

Cubs born in the wild in February appear to be nearly as big as small adult sows by the autumn, but their growth rate varies with the available food and this in turn is affected by weather conditions. During dry summers in Britain, such as 1955 and 1975, growth was much slower and the cubs could easily be distinguished from adults by their much smaller size up to the following spring. Slower growth rate during dry seasons is likely to be due to lack of available earthworms.

Table 10.1 Growth rate of female badgers in the wild

Month	Average weight (kg)	Number of specimens	Weight range	Approx. age in months
April	1.36	3	0.9-1.8	2
May	—			
June	3.63	4	2.7-6.3	4
July	6.80	1		5
August	6.80	1		6
September	—			7
October	—			8
November	8.60	1		9
December	10.40	1		10
January	9.50	3	8.6-9.9	11
February	9.70	9	6.3-11.8	12
March	9.07	1		13
April	8.07	6	8.8-10.4	14
May	8.30	7	6.3-10.4	15
June	8.50	4	6.8-10.9	16

Growth rate of badgers in the wild is difficult to assess accurately, but by taking the average weight of road accident juveniles for each month a general picture was obtained.

Although the sample is small, Table 10.1 shows that there is a gradual increase in weight up to December, but that after January the weights fall off significantly, due presumably to the utilisation of body fat. Lower weights are reached by April, and from then on there is a steady increase once more.

Records for captive badgers show a similar pattern, but the growth rate is often considerably higher. For example, Frank (1940) showed that for his two cubs the male weighed 16.8 kg by 1 January and the female 12.2 kg. So much for plenty of food and comparatively little exercise!

Occasionally in the wild, one cub from a litter may be abnormally small. Such animals seldom live long, but there are records of these runts surviving several weeks after their first emergence above ground.

THE LACTATION PERIOD

A sow will suckle her cubs for at least 12 weeks during a normal year, so in south-west England it is usual for weaning to start in early May. At this time, the cubs are still dependent on the mother for food, but they supplement their diet by foraging near the sett. Usually, they are independent at 15 weeks by which time the permanent dentition is functional and the majority of milk teeth have been replaced.

Exceptionally, the sow will continue to suckle her cubs for a much longer period. During the dry summer of 1975 a sow which had given birth to two cubs, probably in February, was still suckling them at the end of July when they were nearly as big as she was. This was an adaptation to conditions

REPRODUCTION AND DEVELOPMENT

Figure 10.3 Cubs aged five weeks, raised in captivity. In the wild, they would remain underground at this age

when food was extremely difficult for a cub to find. Apparently, the sow was able to keep up her milk supply by ranging further and feeding on different food. As late as the autumn I have also examined dead sows in which the teats were large and the mammary tissue still secreting. This ability to continue suckling during adverse conditions is clearly of great survival value.

The sow nurses her cubs underground, lying on her side or back. When several weeks old, the cubs will knead her mammary glands as they suckle, so towards the end of lactation her abdomen may be almost devoid of hairs.

When the cubs are about 6–8 weeks old, the sow returns to the sett three or four times during the night between bouts of feeding. It is probable that she nurses them on her return on some of these occasions, judging by the time she remains below. At other times she may only come back for a minute or so, presumably to see that all is well with the cubs.

During our all-night watches at Camberley when the cubs were below, the sow's movements were noted as follows (the missing data being due to her using another entrance out of view).

Date	Emergence	Return	
21–22 April	19.56	20.52	GMT
	21.10	21.36	
	23.03	?	
	02.08	03.15	
	03.22	03.40	
27–28 April	20.00	21.01	
	21.02	22.45	
	23.51	?	
	02.23	03.20	

Notini (1948) stated that regurgitation of food occurred during the weaning period, but apart from my seeing cubs licking around a sow's jaws there was no corroborative evidence until Howard and Bradbury (1979) found the gizzards of earthworms and fibrous material in the gut contents of a 1.1 kg cub. Regurgitated food seemed the likely explanation. But on 1 June 1982, Chris Ferris saw regurgitation in the wild at close quarters and I will paraphrase her account:

> The mother left the field edge and came over to her two cubs. I saw something drop from her mouth—a piece of worm which a cub gobbled up. In its desire to get more, it jumped up excitedly at its mother's face and began puppy like to lick around her mouth. She turned slightly away, but her boisterous cub persisted with the licking. She then began to heave, coughed and brought up a great heap of worms which the cub proceeded to eat. Ferris saw her regurgitate again six nights later, this time without being stimulated by the licking of the cub.

Eric Ashby has filmed rather similar behaviour, but circumstances suggested food carrying rather than regurgitation. He writes of a sow feeding in a ditch a few metres from a sett where cubs were playing near the entrance.

> Now and again she noisily ate something, perhaps a snail. Suddenly she left the ditch and visited the cubs. Again she went to the ditch, finding something to eat now and again. Once more she came back to the cubs and this time I noticed her mouth was slightly open and on reaching the cubs she appeared to drop something which was immediately taken by a cub.

A few weeks after weaning, a sow and her cubs may move from the breeding sett. Michael Clark recorded over a period of six years that at one small sett in the three years when cubs were born there, it was abandoned for several weeks in late May or June; but in years when cubs were absent this did not happen until late August or September. After each move, the sett was thoroughly cleared out before being occupied once more. Although large main setts may be occupied continuously throughout the year, there is often a corresponding shift from one part of it to another soon after weaning has taken place.

REPRODUCTION AND DEVELOPMENT

TIME OF SEXUAL MATURITY

Anatomical investigation of the ovaries and uteri of badgers (Neal and Harrison 1958) showed that most female cubs in south-west England become sexually mature when 12–15 months old, but a minority take a few months longer.

Diet plays an important part in determining the time of maturity, as when food is abundant the first oestrus can be earlier. Cubs brought up in captivity and fed well have had their first oestrus in the late autumn—and this is now also known to occur in the wild (see p. 171).

A possible instance of this being followed by mating was reported in the *Naturalist* by John Knight. This concerned a cub born in February 1969 and brought up in captivity. It was allowed free access to the countryside and used to return to its owner each morning. However, by early November it was only coming back about once a fortnight and by the end of the month was assumed to have become fully adapted to the wild. During the winter it was thought to be living in a sett near a garden which it visited on several occasions to forage and raid the dustbin. These visits caused protests from the owner, and eventually an attempt was made to catch it. On the night of 10 April it was watched from the house as it entered the garden, and on being called by the lady who had brought it up, it appeared to recognise her voice. It entered a half-open garage and was shut in. At first the badger was aggressive, but when its past owner and her dog entered the garage its anxiety disappeared and dog and badger sniffed each other amicably. Later on, the badger was examined by Knight who found she was in milk, so she was at once taken back to the sett and within seconds of her entering the hole the young could be heard suckling.

Although it was not possible to prove that the badger caught was the original one brought up as a cub, the circumstantial evidence strongly supports this assumption. If this were so, it would mean that the badger had mated during its first year, probably in the late autumn when 8–9 months old, implantation had occurred after only a short period of delay, and the cubs born in March.

But the most conclusive evidence of early maturity comes from Chris Ferris' remarkable observations of badgers in Kent which she got to know intimately and individually and which became habituated to her presence. One sow had her litter on 7/8 February 1983 and her two cubs were observed regularly throughout the year. In May the female cub injured her paw and became unwell, but she allowed Ferris to examine the wound. This was treated with hydrogen peroxide and antibiotics were given in food. By mid-June the cub had quite recovered. From then on she was completely trusting and would greet Ferris whenever she appeared. In December she was seen to mate three nights in succession on 3/4/5 of that month. She was then seen regularly, but on 24 January only briefly. The next night she did not appear at all, but on the 26th she came to greet Ferris who noticed that she smelt of milk. She confirmed that she was indeed lactating. So her cubs were born on the night/morning of 24/25 January giving a gestation period of seven weeks plus, presumably with no delay in implantation (see p. 175).

Normally, a sow is mated when a year old, or soon after, and has her first litter when two years old. I had proof of this when on 26 December 1945, some badgers were dug out and the two adults were killed, but the female cub, aged about ten months, was marked on her ear and released. She was caught again at the same sett on 16 June 1947, this time with cubs of her own, aged about four months. So the mother was born in early spring 1945, mated in 1946 and had cubs in 1947.

Anatomical research on male badgers by Ahnlund (1980) in Sweden and Graf and Wandeler (1982) in Switzerland showed that the majority became mature in the spring and summer of their second year (12–15 months) although some may take longer and very exceptionally may not be mature until two years old.

DISPERSAL OF THE CUBS

The time when the young leave the parental sett varies according to circumstances. A few may leave in the autumn, more the following spring or summer; some remain permanently.

Telemetry studies by Cheeseman in a high density area in the Cotswolds showed remarkably little evidence of yearlings emigrating. He marked and released 629 badgers and of those recaptured in subsequent years only seven males and one female had changed social groups. Much appears to depend upon whether the composition of the social group allows the yearlings to remain. Females normally stay, males may emigrate on nearing maturity.

Burness (1970a) found that his male albino left its parental sett at 16 months, but moved only 0.4 km. It mated with a young sow which subsequently had cubs. During nine years the albino was never known to return to the parental sett, but was based for varying periods on three setts within an area of about 200 ha.

REGULARITY OF BREEDING

For the southern half of Britain we found that nearly all adult sows examined between July and the end of the year had either blastocysts or embryos (Neal and Harrison 1958). Similar findings have since been made for France, Sweden and Switzerland. Ahnlund (1980) found that in Sweden 90–95 per cent of sows two years or older had blastocysts, but yearlings had fewer (66 per cent). However, at post-implantation stage the proportion of sows bearing fetuses may be much less than at the blastocyst stage. This reduction appears to vary with population density. In Sweden where density is low Ahnlund found there was no significant drop at implantation stage, but from high density areas in south-west England the reduction was as much as 50 per cent or more (Anderson and Trewhella 1985).

Further evidence for the variability of this reduction with density was obtained by estimating the proportion of sows in a sample found to be lactating—a good indication of successful breeding. From 71 sows examined by us between 1948–70 during the probable lactating period February–May inclusive 34 (48 per cent) were lactating and 37 (52 per cent) were not. But of the latter 29 were in their second year and too young to have litters although the majority were mature and many had mated. So, only eight (18 per cent) of possible breeders were not lactating in this sample from Somerset, a region of moderately high density. By contrast, in a Ministry sample from very high density areas of south-west England the proportion of sows lactating was only 32 per cent (Gallagher and Nelson 1979). However, some of this difference could be accounted for by early loss of complete litters due to starvation, predation or accident. Kruuk (1978) suggests that typically only one litter is produced per social group although other sows may be present, due to the killing of other cubs by the dominant sow. But evidence for this was from a captive colony comprised of badgers derived from diverse sources. It is not known whether this happens in the wild; on the other hand there is much evidence that more than one litter per group may be raised successfully.

There is no doubt that individual sows can breed regularly. G. Dangerfield kept one sow which had litters four years out of five and observations in the field on easily-identified animals by R.W. Howard and others confirm this.

TIME OF MATING

There has been much diversity of opinion as to when mating takes place, each month having been suggested by one authority or another.

Fischer's embryological data (1931) strongly suggested July. I agreed with Fischer's conclusions after seeing mating take place three times in July in Gloucestershire. This seemed to prove it, but my conclusions were premature. Not long afterwards R.W. Howard saw matings in April, so the season seemed not only to be a long one, but appeared to be stretching earlier and earlier.

It was therefore imperative that intensive watches should be carried out during the early months of the year, to see what was happening then. My son Keith and Roger Avery planned to do this, and succeeded over the three winters of 1954-6 in watching 13 times in January, ten in February and 28 times in March, and on many more occasions in April and May. Their efforts provided much useful data and demonstrated without doubt that February was the month of greatest sexual excitement.

More recently, badger watching has become very popular and many more accounts of mating have been described. Of these, those by Paget and Middleton (1974a) are particularly valuable as they, too, largely concentrated on the early months—a time avoided by the less hardy watchers!

Mating was also observed by Chris Ferris on 35 occasions during five years of intensive observations between 1979-83. Because the badgers had become habituated to her presence she could watch them from close quarters away from the sett area. This she did throughout the year in all weathers. Her eight observed December long-duration matings comprise a unique record as previously no work of such an intensive nature had been carried out during this month in such a manner. What is more, the December matings were all estimated to involve sows under a year old. This may be very exceptional in the wild as all these records are from an area of Kent where market gardening is carried out and food is plentiful due to regular irrigation and hence the greater availability of earthworms during dry spells (p. 169). This is comparable to cubs in captivity having large quantities of food.

The duration of copulation is also very variable. Many instances reported were for less than two minutes, but usually they have ranged from 10-90 minutes. Some of the short-duration matings recorded were probably only rutting behaviour where penetration was not attained: these can be ignored, but others certainly appeared to be definite short copulations. I believe some of the latter may coincide with a secondary oestrus (p. 176).

REPRODUCTION AND DEVELOPMENT

Long-duration pairing is characteristic of other members of the *Mustelidae* such as the mink, *Mustela vison* (30–40 min.), the sable, *Martes zibellina* (50 min.) and the polecat, *Mustela putorius* (up to three hours).

When long-duration matings are plotted (Figure 10.4) they appear to be well spread out although the main peaks occur in February and April/May. It should be borne in mind that far more badger watching has occurred between April and September than in winter, and I am sure that if more people had watched in February the peak for that month would have been much higher. This supposition is strongly confirmed by anatomical investigation of testis activity (Neal and Harrison 1958; Ahnlund 1980; Graf and Wandeler 1982). These have shown that testis weights and sperm production are greatest during the period February–May, but animals with sperm in the tubules of the epididymis may be found in any month making successful mating possible throughout the whole year.

The February peak for mating is certainly correlated with a post-parturient ovulation occuring either a few days or, at most, a few weeks after the birth of the cubs. This appears to be a normal pattern in south-west France (Canivenc and Bonnin-Laffargue 1966) but in Britain our anatomical evidence suggests that it occurs frequently, but not always.

Figure 10.4 Frequency (in half months) of long-duration matings

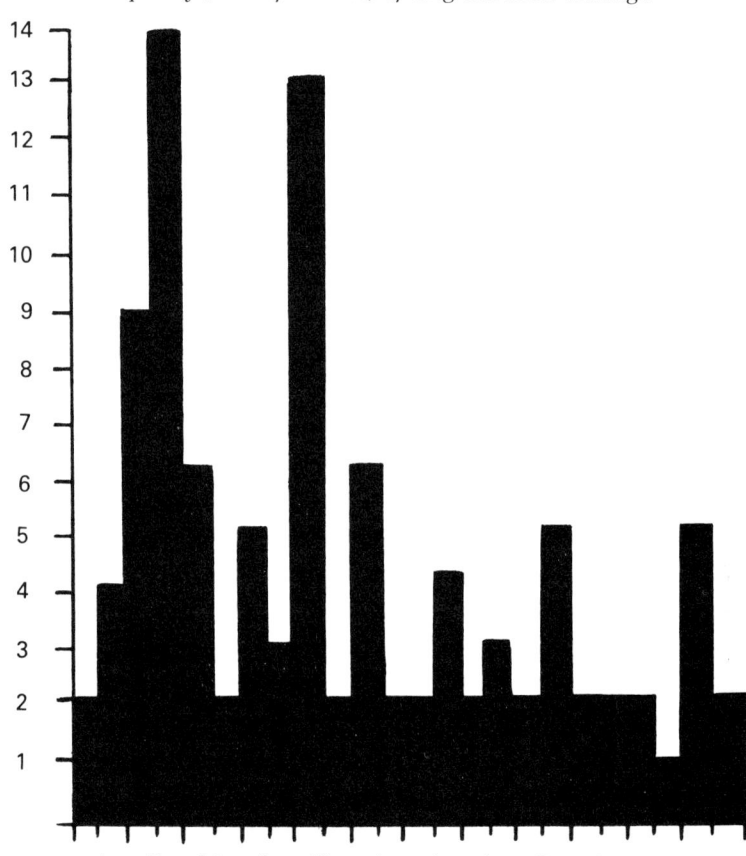

The second peak in April/early May could well be correlated with two possible events. Yearling sows often come into their first oestrus about this time, and in early May in particular, lactation is slackening off and this may trigger off another oestrus.

To make the picture even more complex mating is dependent upon an oestrus of only a few days duration and the onset of oestrus is subject to factors which are variable. Oestrus closely follows parturition in many sows, and that event can vary. The time of cessation of lactation is also variable as is the first oestrus of yearling sows. So these variable factors probably account for long-duration matings seen during other than the peak months.

Further aspects of the problem will be discussed under delayed implantation (p. 175).

MATING BEHAVIOUR

Mating behaviour varies considerably both in duration and intensity. With some pairings no preliminary play of any kind was seen but with others there was considerable excitement with violent scuffles and much emission of scent. Usually, the female received the male quite passively and was not observed to initiate the mating.

The first sign of interest shown by the boar is often the raising of the tail into a vertical position and the emission of a loud and often continuous deep whinnying purr. The boar may then approach with a shuffling motion, taking short steps with the legs kept rather rigid. On mounting, dog-fashion, the boar grasps the sow with his front legs in front of her flanks. He keeps his position by gripping her neck or ear with his teeth, sometimes weaving his head from side to side before getting a grip. The sow may give a yelp if the bite is too vigorous. Pairing has been seen both when cubs have been present and when adults have been living alone.

There is no doubt that long-duration matings, ranging from ten minutes to an hour or so, occur when the sow is in full oestrus. At this time there is great excitement among the males and some rivalry may be apparent if several males are present.

The length of oestrus is probably 4–6 days. Howard recorded that over the period 1–6 April at one sett he saw only rutting behaviour on 1 April, on 4th and 5th copulation occurred on and off for several hours, but on the 6th all signs of sexual activity had abated. Paget and Middleton saw a very similar sequence of events in Yorkshire. In captive badgers, John Sankey saw prolonged mating on five consecutive days and nights (4–8 May) and in another year there were three days of long-duration matings (22–24 May) preceded by nights when shorter pairings occurred. Eunice Overend told me that her tame sow when on heat became extremely restless for about ten days. It was possible to see her swollen vulva which was pink when in full oestrus, as in bitches.

So when a sow is in oestrus she will be mated for long periods, on and off for several days and nights, and as ovulation is induced by the stimulation of copulation this pattern of events may be important in bringing about fertilisation.

In February, many sows have a post-parturient oestrus and on these nights the boar (or boars) is very excited. At this time he is kept away from the region of the sow's nest chamber and lives elsewhere. Usually he emerges first and trots from hole to hole and over the whole area of the sett making deep vibrant whinnying purrs almost continuously. He will sometimes half-disappear down a hole but will not enter, presumably because the sow below would be aggressive. His whinnying suggests that he is trying to call up the sow, but the wider patrolling of the surroundings may be to keep the area free from potential rivals. On such nights the sow may not emerge at all, but if she does, mating usually follows.

Later in the season, the pattern of behaviour is rather different as the cubs are larger and the sow no longer has to defend them. In April and May both boar and sow may emerge within a short time of each other and mating may take place almost at once. On many occasions prior to mating, the boar was seen by Howard (1951) to turn his back on the sow, raise his tail and back on to her, setting scent on her body. During copulation there is much vocalisation. Drabble (1970) fixed a microphone in the roof of an artificial sett in the grounds of his home in Staffordshire and on 4 April succeeded in recording the vocalisation made during what was almost certainly a mating of about 20 minutes. This ended in a loud growling threat display.

Paget and Middleton (1974b) described how in early June, the sow emerged 20 minutes before sunset, followed almost immediately by the boar. There was a period of scratching, mutual grooming and musking and the boar started to 'purr' loudly. He then began to paw the ground before mounting the sow. After mating for about 25 minutes, the sow ran off, but the boar 'spent further minutes both sniffing the ground where the mating had taken place and then scratching soil over it before leaving hurriedly after the sow'.

On nights when sexual excitement is considerable, violent play or attempts at mating may suddenly be interrupted by a bout of violent digging or the bringing back of a bundle of bedding towards the entrance, only to be abandoned. This appears to be displacement activity.

In large groups there may be several boars and when a sow is in oestrus this may attract others. Badgers can certainly be promiscuous. Howard observed two boars copulating with the same sow during the evenings of 4 and 5 April 1951. It started when two boars which were purring loudly literally dragged a sow out of an entrance. Both boars tried to mount the sow and one eventually gained control and remained in the mating position for some time, while the second boar stood by 'purring' loudly. A sudden noise caused all three to dash back into the sett, but after about 20 minutes they re-emerged and one boar immediately mounted the sow and for the next hour was in the mating position. Meanwhile, the second boar wandered about the sett digging little holes here, there and everywhere and occasionally approaching within a few feet of the copulating boar, but on receiving a threatening look and a loud angry 'purr' he would retreat. Sometimes the attack succeeded and the mounted boar would dismount and chase the other away. When this happened, it was clear that the sow wanted attention as she would run up to the struggling boars and given the slightest chance would raise her tail and back on to the nearest boar's flank. The boar would then raise his tail and mount. The next night one of the boars mated with the sow for 40 minutes but dismounted to investigate a strange badger which came near the sett. Meanwhile, the second resident boar took the opportunity of mounting the sow and remained in undisputed possession for a further 40 minutes. The first boar on returning, no longer took any interest in his rival and allowed mating to continue. At times when neither boar was

REPRODUCTION AND DEVELOPMENT

mounted, one would guard her and if the other approached would go to the sow, smell her, turn round, raise his tail and set scent on her. As Howard remarked 'the affair was one, not of billing and cooing, but of tailing and purring'. Paget and Middleton mention another instance in Yorkshire in June 1972.

Mating has been recorded mainly around the sett following emergence, but Chris Ferris has also seen it many times away from the sett, both in open fields and in woodland.

Do they mate underground? This is by no means impossible in well-established setts, as some of the tunnels and chambers can be large enough for mating to take place. All that can be said at present is that it certainly does happen in artificial setts where there is plenty of room.

DELAYED IMPLANTATION

One of the remarkable things about the reproductive cycle of the badger is that mating and fertilisation can take place during any month of the year, but that cubs are normally born at one season only. This is brought about by delayed implantation, a phenomenon first shown to occur in the badger by Fischer (1931). When this occurs, the blastocysts on reaching the uterus do not implant in the usual way, but remain free within the cavity of the uterus for a long time before becoming embedded in the uterine wall. During this free state, the blastocysts keep alive on a minimum of oxygen and food that diffuses through the wall, growing very little.

Figure 10.5 Photomicrograph of a blastocyst recovered from the uterus in June. Diameter 1.5 mm

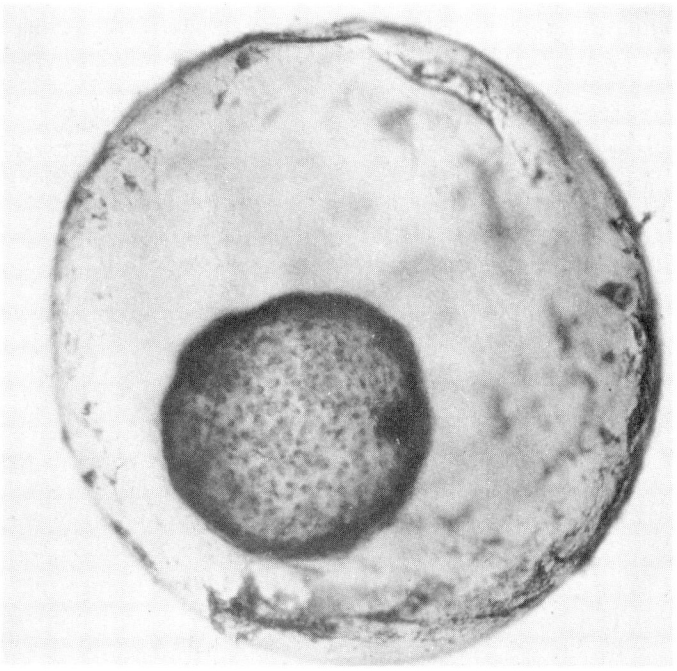

One of the strangest phenomena associated with delayed implantation in the badger is that sows may sometimes have one or more further oestrus periods during the course of delay, even though live blastocysts are present (Neal and Harrison 1958). During these secondary oestrus periods, she will accept the male for a further mating, but in my opinion almost certainly for a short duration only. However, the stimulus of copulation appears to be enough to cause a further ovulation and the formation of additional corpora lutea, although these eggs are not fertilised. I believe this is the explanation of short-duration matings seen in the field and the increase in the number of corpora which occurs towards autumn. (In our investigations we found the average number of corpora between February and June was 4.2, but from July–December was 6.2.)

Further ovulations during delay have since been confirmed in both Swedish and Swiss badgers (Ahnlund 1980; Wandeler and Graf 1982). Also, Richard Paget (1980) showed that a Yorkshire sow had at least two and possibly three periods of oestrus during one year.

It is impossible to know the exact time of implantation in any one individual, but the average time for the badgers in one region can be calculated from the dates of birth recorded, as the period from implantation to birth is seven weeks, plus or minus a few days. Hence, for south-west England the average implantation date is late December.

Reference to the average birth times in different latitudes (p. 162) shows that for south-west France the average implantation time is mid-December, in south-west England it is late December, and in the north of England, early January; in Scotland, Sweden and Germany, mid-January and in parts of Russia late January or early February. Thus, if fertilisation takes place as a result of a mating soon after the birth of the cubs, the period of delay may be around ten months, but if mating occurs later in the year it will be correspondingly shorter. So with an October mating, for example, it may only be around two months (Figure 10.6). However, the range of birth dates in any region is considerable, so the variation of implantation times must be correspondingly wide. In Britain, it ranges from early November to February. It therefore follows that the full period of gestation, from fertilisation to birth, varies from as short as four months to nearly a year.

Exceptionally, implantation can take place without a period of delay. Evidence for this comes from Chris Ferris's records of one early November and eight December matings, almost all certainly involving sows under a year old. For one of these she knew the exact age and gestation period. However, it must not be assumed that all such matings with young sows lead to fertilisation and implantation.

One of the intriguing problems of the reproductive cycle is what factors cause implantation to take place at the appropriate time. There is little doubt that one factor is the influence of oestrogen and progesterone on the lining of the uterus, making it spongy and receptive to the blastocyst. But what causes these hormones to be secreted in the correct proportion at the appropriate time?

It is known that variation in day length influences the time of implantation in some mustelids (Pearson and Enders 1944) and although badgers are strictly nocturnal in winter, and consequently unlikely to be affected by this factor, it has been suggested by Canivenc and Bonnin-Laffargue (1966) that increase in night length correlated with greater activity may be factors.

But if activity is a factor it should apply equally well to badgers in all

REPRODUCTION AND DEVELOPMENT

Figure 10.6 Examples of variation in the period of delayed implantation in the badger, caused by February, May and autumn matings. I =implantation, P =parturition (birth), M =mating, against the month of the year

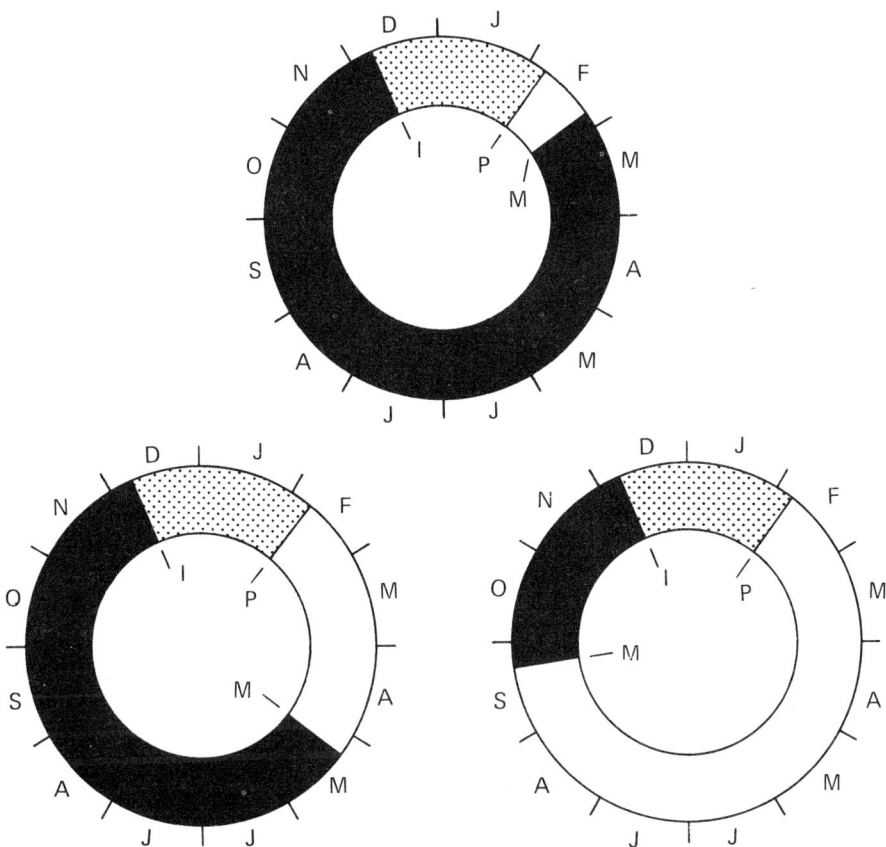

regions. But in northern Russia, the badger, like the bear, is in a semi-dormant state during the winter, bedding up completely by the end of November and living on its stored fat. A short warm spell in cold winters does not bring them out, but in mild ones some may emerge in January or February (Novikov 1956).

By contrast, in south-west France, the milder winters may allow badgers to feed normally, and activity may be a usual thing. In south-west England, although some activity occurs all through the winter, except during periods of severe frost, fewer animals come above ground regularly. Where regular counts have been taken at main breeding setts the numbers seen have gone down dramatically around the middle of November. From then to late December (implantation time) there has been less activity than at any other time of the year.

This drop in numbers in November could be due to dispersal, but another possibility is that most of the original number are there all the time but some are lethargic and do not often emerge. Evidence is accumulating that this is indeed the correct explanation and that it is largely the younger animals that are more active during the month of December.

Those who have kept badgers in captivity have observed that after stuffing themselves with food in the autumn, they then lose interest, sleep a great deal and do not become active again until after Christmas—a period of semi-dormancy of several weeks.

It is certainly my belief that implantation takes place during this inactive period and that any marked activity at the end of December is more characteristic of the post-implantation period.

Breeding records in captivity also suggest that only when badgers are relatively undisturbed over the implantation period is success regularly attained. Sankey found that although long-duration mating occurred on four successive years, litters were only born during the year they were left undisturbed during the December period.

Further indirect evidence for this comes from Canivenc's own investigations. Of the 752 female badger carcasses he examined only five were obtained during the implantation period; this reflects the difficulty in obtaining badgers at this time, presumably because they are then semi-dormant below ground.

There are some significant parallels here with other mammals. Bears have delayed implantation too. Like the badgers, autumn for them is the time when their main occupation is feeding and they put on much fat. Food then becomes scarce and they bed up. They become lethargic, but like the badger their temperature does not drop. It is during this semi-dormant period that implantation takes place (Winsatt 1963).

Increase in night length seems to be a most improbable factor for implantation, but Canivenc, after a number of years of research, succeeded in 1971 in inducing early implantation. Using six captive sows which had been mated soon after giving birth to cubs, he placed them in an experimental chamber in May, where they were subjected artificially to a gradual increase in night length and a reduction in temperature of 10°C. Implantation occurred in July.

This experiment proved conclusively that under artificial conditions increase in night length, combined with a drop in temperature, brought about implantation. But is it possible for these factors to operate under wild conditions? I cannot agree that they can, for the following reasons. First, badgers are strictly nocturnal after the end of October and thereafter their times of emergence are well after dark and in no way related to sunset times (see Figure 7.1). Second, as night length increases more dramatically in northern latitudes and temperatures drop more severely, you would expect that implantation would be earlier in northern countries. The opposite is in fact the case (p. 176). Third, the range of dates for implantation in any one region may be very wide, some being well before the shortest day, others very much later. In Britain, there are exceptional records as early as October and as late as May, although the majority occur in late December and early January. Lastly, there is the phenomenon of two sows living in the same sett under the same conditions of night length and temperature, but having their cubs as much as five weeks apart. This discrepancy is not unusual.

When looking for a factor which triggers off implantation, it is logical to seek for one which applies equally well to badgers in the sub-arctic as in southern France. This factor must also be variable, as it not only has to account for the variation due to latitudes but also to wide variations in any one locality. This suggests that a factor other than a climatic one is at work—one which is related to the characteristics of each individual.

I believe that relative inactivity is the key to the problem, as it is during the state of semi-dormancy that implantation occurs. It is well known that the putting on of much fat is a pre-requisite for hibernation or semi-dormancy in many species, and I believe this is undoubtedly so in the badger. But the putting on of fat varies with the food supply and the time available for foraging, and this could account for the variability in implantation times. There is also evidence (p. 186) that after dry summers and autumns when feeding has been very difficult, either a poor year for cubs follows (possibly due to implantation failure) or the litters are later than usual. This could be explained by this hypothesis.

But why should semi-dormancy bring about implantation? I would suggest, very tentatively, that the explanation could be as follows. It is certain that one factor causing implantation is the secretion of steroid hormones which cause the uterine lining to become receptive. In most mammals this occurs soon after fertilisation when oestrogen and progesterone are secreted from the ovarian tissues. In the badger, this occurs after a long period of delay and during the latter, only very small amounts of these steroids appear to be secreted—not enough to bring about implantation. However, steroids are soluble in fat and I believe that as a badger puts on fat these substances become stored up in it in the same way that insecticides such as DDT are known to be stored. This continues until the sow goes into a state of semi-dormancy. It is then that she lives on her stored fat and as this becomes diminished, the stored steroids are released and shed into the blood stream in sufficient quantities to bring about implantation. Much more research is needed before this hypothesis can be accepted or refuted, but it does provide a possible explanation of the variability in implantation times both among individuals in the same locality and according to latitude, which an external factor such as night length fails to do.

One further factor which should be discussed is the effect of stress. This is known to retard implantation in rats. Badgers captured as adults and subsequently kept in isolation have on occasions produced cubs up to 15 months afterwards (Cocks 1903), so it is likely that the stress caused by being kept in captivity delayed implantation beyond the usual time.

Delayed implantation is a remarkable adaptation to ensure that cubs are born at the most appropriate time for survival. This, according to the restrictions of climate, is as early in the year as possible. In spite of wintry conditions, the new-born cubs in January and February are well provided for. They are kept warm, thanks to the central heating system provided by the mother and the cubs themselves and the insulating properties of all the bedding in which they lie. Because of the long period of lactation, the cubs are already large by the time they are weaned and the season is then sufficiently advanced to provide plenty of food. This gives them the whole period of summer and autumn to put on enough fat for survival during the severe winters which are characteristic over much of the badger's range. It would also appear that delayed implantation has enabled southern species to disperse into colder regions.

How the delay period has come about is a matter of speculation. However, it would seem to be significant that within the *Mustelidae*, those species which have more than one litter a year never show a delay, but the majority of those with one litter do.

Is it possible that the badger's ancestors were smaller animals which had more than one litter a year and that during the course of evolution a second litter became disadvantageous? If this were so, it could account for the further oestrus periods which take place in many present-day sows even though healthy blastocysts are present. Perhaps, too, the retention of these further oestrus periods in the cycle is a safety device to ensure eventual fertilisation even if earlier matings have been unsuccessful?

Alternatively, perhaps the badger's ancestors were at one time polyoestrus with cycles every 4–6 weeks, between February and October, and that these cycles became partially or completely suppressed with the supervention of delayed implantation?

Or, as Ahnlund (1980) suggests, matings throughout the year including during the delay period may serve to strengthen the bond between the sexes thus helping to maintain the male's defence of the social group's territory over a longer period.

11 Numbers

HISTORICAL

It is not possible to assess changes in the badger population in Britain with any accuracy. However, from Parish Council registers in the seventeenth and eighteenth centuries, there are many records of head money being paid for badgers. They come from many parts of the country and suggest that the badger had a wide distribution in those days. Payment ranged from 4d to 2 shillings per head.

There was considerable persecution of the badger in the nineteenth century and numbers dropped considerably. By the end of the century, badgers were considered rather rare, but this may have been due to paucity of records as most estimates were based only on local experience.

Lack of keepering during the period of World War I, and for some years after, gave some respite to the badger and numbers steadily increased, so that by the early 1930s they were fairly common in suitable areas all over England and Wales, but this increase was less noticeable in Scotland.

Frances Pitt (1935) reckoned for the Wheatland district of Shropshire that the number of setts had increased from ten to 37 in an area of 220 km^2 in the 34 years from 1900. This would seem to be typical of the general rate of increase in the country as a whole.

During and after World War II, the population continued to increase steadily up to about 1960. No doubt this was helped by the increasing tolerance towards the badger by landowners, farmers and those concerned with hunting and shooting, the helpful policy adopted by the Forestry Commission who welcomed badgers on their land, and a more favourable public opinion generally. During this time, too, there was a significant increase in the number of nature reserves on which badgers were protected.

However, between 1960 and 1972 a number of adverse factors were increasingly taking their toll and in some areas numbers started to decline. These were the ever-growing number of badgers killed on the roads and railways, the possible reduction in fertility due to much greater use of pesticides of the chlorinated hydrocarbon type, a great increase in some parts of the country of badger digging and the use of cyanide gassing on farms and shooting estates. The latter was illegal, but was nevertheless practised as a routine measure in many places. The effect of these adverse factors was very much greater in some districts than others so although badger numbers continued to rise in some regions they fell quite dramatically in others.

The passing of the Badgers Act in 1973 resulted in a reduction in badger digging for sport and this allowed another increase in the population in areas particularly affected by persecution.

By 1974, there is little doubt that in the southern half of Britain population density was higher than at any previous time. However, in the last decade numbers have been significantly reduced particularly in parts of the Midlands, South Wales, Yorkshire and some northern counties due to a considerable increase in illegal digging. There has also been a drastic fall in numbers in those parts of south-west England where culling took place due to the bovine tuberculosis problem. In Scotland, there are indications that numbers are continuing to increase steadily.

DISTRIBUTION OF SETTS

Setts have been mapped on a county basis in Britain by those taking part in the National Badger Survey organised by the Mammal Society. Data was also received from the Biological Records Centre and the Forestry Commission. From these sources nearly 20,000 setts have been located and for 12,000 habitat details are known (p. 68). Using these data as a base it was possible to guess, within wide margins, the number of setts which were probably present in many counties. This was done by taking the actual number of setts already located in a county and modifying this figure according to the type of geology and land use of the parts not surveyed adequately. For some counties which have been extremely well surveyed, the estimate is reliable, but for some areas, especially in parts of Wales and Scotland, where little work has been done, the results must be treated with great caution. Nevertheless, as a rough approximation I'm sure the exercise was well worth doing. Great credit for this should go to E.D. Clements, who took enormous trouble in sifting and correlating all the data, and for drawing the map.

This map (Figure 11.1) is based on sett densities in 10×10 km squares. For such large areas it should be stressed that in Britain almost all these squares show great diversity of land use within their boundaries. Hence, in many cases only small areas provide optimum conditions for badgers and these are typically surrounded by less favourable regions.

In exceptionally favourable districts for badgers, the survey showed sett densities of over 100 per 10 km^2. In more typical areas which may be described as good badger country, 50 setts per 10 km^2 are often found.

Extraordinarily high numbers of setts have been found in a few regions, due perhaps to a local population explosion of badgers and excellent digging conditions. For example, in 1971 officials of the Ministry of Agriculture mapped 443 setts in an area of just over 10 km^2 (40 sq m) near Thornbury in Gloucestershire and in a much smaller region in Dorset, 247 setts were found in 1,217 ha (4.7 sq m). It is obvious that a high proportion of these setts were not occupied at any one time and many were small, but clearly the population was very high indeed. Both these high density areas were associated with local incidence of bovine tuberculosis in the badger population (p. 201).

Figure 11.1 Distribution of badger setts (occupied and unoccupied) in England, Scotland and Wales (absent from Orkney and Shetland) with reliability diagram inset. After E.D. Clements

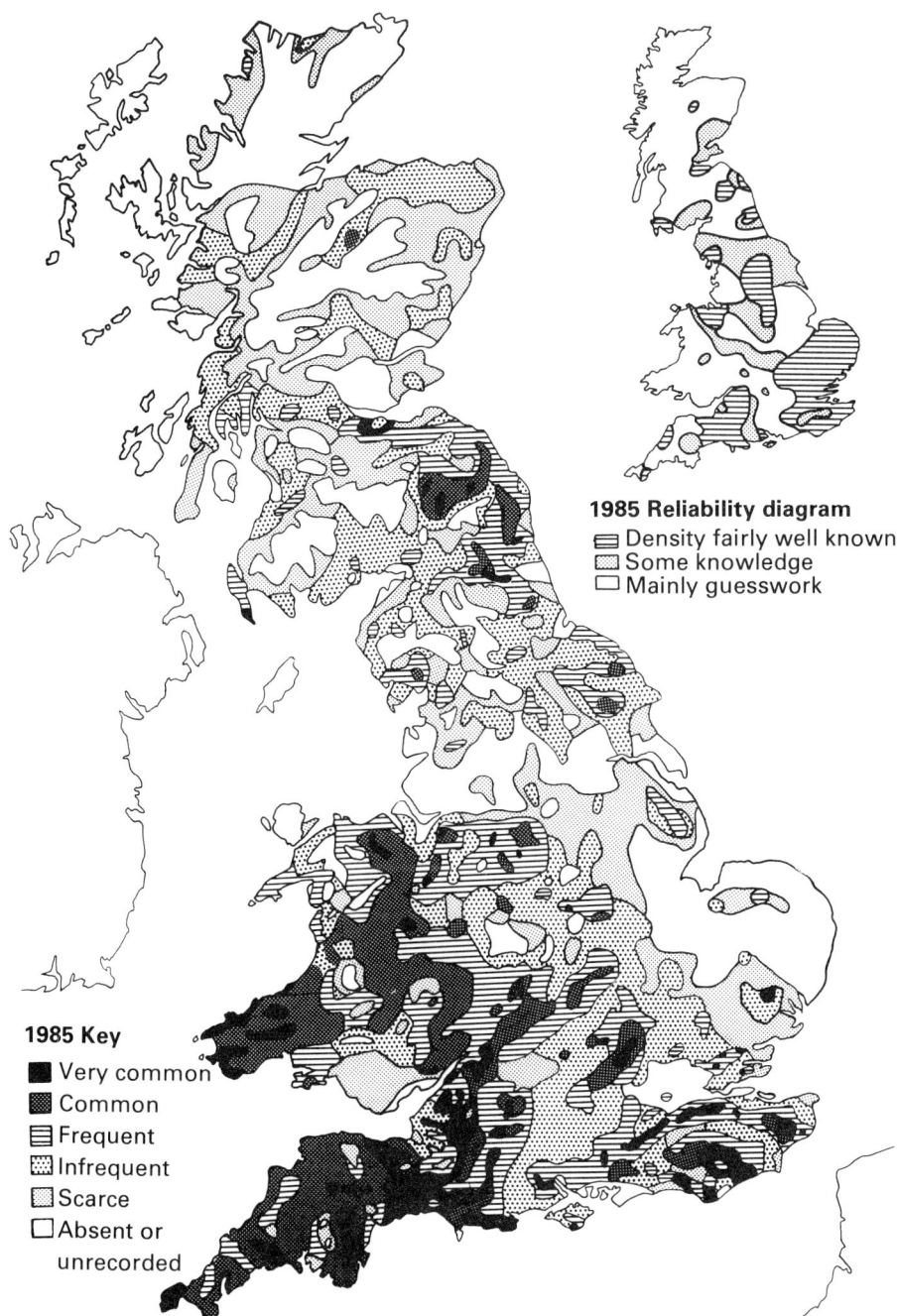

Studies by Cheeseman and Mallinson (1981) and Kruuk and Parish (1982) showed that although social group size varied according to local

Table 11.1 Territory size, group size and badger density in four areas (n=number of social groups in sample). By courtesy of Cheeseman and Mallinson (1981)

Area	Mean territory size (ha)	Mean group size	Mean group size (adults only)	Density (no. per km^2)	Density (adults only) (no. per km^2)
Glos 1 (n=6)	22.0	4.8	4.3	22.0	19.7
Cornwall (n=6)	74.7	4.8	3.3	6.5	4.7
Avon (n=7)	73.7	5.7	3.6	5.8	4.9
Glos 2 (n=5)	24.8	7.6	5.8	30.7	19.4

conditions, the population density was not correlated with group size; it was the size of the territories that was significant. Thus in very favourable country, territory size was small and density high and vice versa.

Cheeseman's data from operations where whole social groups were removed (Table 11.1) showed that for those in Cornwall the density was 6.5 adults per km^2, in Avon 5.8, and in one valley in the Cotswolds 20. Kruuk's data from three very different areas of Scotland showed 3.5 per km^2 (range 1.1–8.2).

Any reliable calculation of population density for large regions of country can only be based on the number of social groups present. If their number was indicated by the total number of setts present, these could be counted, but unfortunately there is no such correlation. However, if only occupied main setts were taken into account a more realistic figure could be arrived at, but this would only apply to areas where interference and persecution were minimal. For example, it would not apply to parts of Yorkshire, South Wales and some northern counties which have a much lower density than the number of setts suggest because much digging has occurred.

The other problem is determining which are the main setts. In many regions this presents little difficulty as they are large, well spread out and used all the time, but where sites within a territory are favourable subsidiary setts may be dug which can be quite large and used for most of the year, particularly at the breeding season. These may look like main setts, but belong to the same social group. So Figure 11.1 should be interpreted with caution as setts other than main setts are included. However, it is of value in demonstrating the suitability of different parts of the country for badgers and where there is no significant interference, it is a useful rough indication of relative badger numbers.

FACTORS INFLUENCING POPULATION DENSITY

There is good evidence that in many high-density areas in Britain the number of main setts occupied has remained fairly constant over several decades if human interference has been minimal. And it can be assumed that apart from periodic fluctuations in numbers the population too has remained steady.

When population density is steady, it follows that recruitment into the population through births and immigration must equal the loss through deaths and dispersal. So any change in density which occurs will be the result of factors influencing any of the four items in the equation.

In order to find out the average recruitment of cubs into a population per year we must take into account the sex ratio, age of maturity, proportion of breeding females, average litter size and life expectancy.

The sex ratio at birth is 1:1. Figures from road casualties obtained from the Ministry of Agriculture for south-west England, combined with my own, showed that out of 571, 278 were males and 283 females. However, Cheeseman found that the ratio of adults at breeding age showed a marked preponderance of females. The discrepancy between the data may be due to the greater likelihood of males becoming road casualties, since they are more widely ranging.

Sows and boars usually become mature when 12–15 months old (range 9–24). However, although most sows mate, not all bear litters. So although the average litter at birth is 2.7 the average productivity per female per year is only 1.67 and the figure for birth rate per head of population is 0.6 (Anderson and Trewhella 1985).

Life expectancy from birth as calculated from age distribution within populations is between two and three years. However, this may be an overestimate as this method takes no account of deaths underground during the first eight weeks.

It is not known how long a badger can live in the wild, but Chris Ferris reports a boar, known well since a cub from a large mark on its flank—an old traffic injury—which was killed by diggers at $11\frac{1}{2}$ years. She described it as 'big and rather ponderous, but hale and well; it had sired 3 cubs the year it died'. I once saw a badger of great age, it was round as a barrel and very stiff in the joints, and there are several records of other very ancient animals being dug out; some of these were blind. However, it appears to be unusual for badgers to exceed ten years, but a very few may reach 15 or even more. A captive badger kept by Ruth Murray reached $19\frac{1}{2}$ years.

The mortality rate is difficult to assess accurately, but the most reliable method is based on age distribution in the population. However, the accuracy of this method is dependent upon each age class being sampled equally and this is difficult to achieve. The regular trapping, marking and releasing of a significant proportion of a population has given good results (Cheeseman unpub.). It is clear that greatest mortality occurs in the first year after which mortality rate is remarkably constant at about 25 per cent per annum. In the first year mortality may be as high as 50–70 per cent. Attempts have been made to estimate losses during the eight weeks following birth, but as this period is spent underground accurate information is difficult to obtain. Comparing size of litters seen above ground with those of birth rates I estimate an 18 per cent loss. Other data based on the proportion of females with fresh placental scars found not to be rearing cubs suggest up to 25 per cent (Wandeler and Graf 1982; Ahnlund unpub.).

It is also probable that when yearlings are dispersing and attempting to join a new social group, rather more losses than usual may occur.

For a study in depth of the population dynamics of the badger, the reader is referred to the excellent paper by Anderson and Trewhella (1985). In this they calculate from fecundity and maturity that the average intrinsic growth rate of the population is 0.2 per year. This slow potential increase is of course regulated by various constraints such as disease or adverse feeding conditions such as droughts. Such adverse factors may cause a population to decline by 30 per cent or more, but recovery is usually achieved in 2–4 years. If, however, a number of social groups are removed, as in disease operations, recolonisation may be very slow. Two removals in Gloucestershire in a high density area have still not recovered to previous levels after six years. Presumably time of recovery varies with the status of the surroundings and the extent of the area where eradication occurred.

It appears likely that in the earlier stages of recolonisation the individuals present do not provide the stability needed within a social group to allow successful breeding. Cheeseman found in one such area that cubs were not produced for three years although adults of both sexes were present.

A possible explanation of this could be that because social groups are such closely bound entities, when a vacuum is produced, recolonisation does not necessarily take place from contiguous groups but builds up in a haphazard manner from wandering individuals seeking either new territories or acceptance by other social groups. If this is so the speed of recolonisation would be governed by the reproductive success of the various social groups in the region and the extent of the area where eradication occurred.

Dispersal of young badgers (p. 170) is extremely restricted in high-density regions. In other areas it would appear that there is a tendency for female yearlings to remain and males to disperse. But how far dispersal plays a part in regulating populations is unknown.

INTRINSIC FACTORS CAUSING POPULATION CONTROL

It is well known that some years are better for cubs than others and it seems likely that any intrinsic constraints on breeding occur at implantation.

If it is true that putting on fat is a requisite before implantation takes place (p. 179) then a summer and autumn when food is very scarce might to conditions when implantation becomes less likely, so fewer litters would be born the following spring. In 1975, badgers in many parts of Britain were in very poor condition due to the severe drought that lasted from June into the late autumn. This also co-incided with a poor acorn and blackberry year. So it was most unlikely that much fat was stored. The spring that followed was certainly a very poor season for cubs in many parts. But whether this was due to failure to put on enough fat is an open question. Obviously, much more evidence is needed before this hypothesis can be proved or refuted.

Failure to implant may also be due to stress. Badgers in captivity, captured as adults, could be considered to be living under conditions of mild stress and it is in such animals that record periods of delay have been noted. It is also true that although mating has occurred many times in badgers kept in captivity, only a small proportion have had litters. So stress may interfere with implantation.

It may be that under conditions of very high density badgers are subjected to increasing stress and in some seasons, difficulty in obtaining sufficient food. These two factors together could play a part in population control when

CAUSES OF DEATH

Natural Enemies

It would appear that badger numbers are not much affected by natural enemies apart from man. Adult badgers in particular are seldom killed by other animals, although there are isolated examples involving foxes (p. 81), and occasionally badgers may be killed by hounds if found above ground. They may also die of bite wounds received during territorial disputes (p. 150).

Cubs are rather more vulnerable. During the first few weeks of life they are left on their own for long periods and a few may be killed by dogs entering the sett during the day. Occasionally, a vixen or a boar badger may kill them. It is also possible that if a sow is disturbed soon after the birth of the cubs she may kill and eat them. This has occurred in captivity.

In February, it is not unusual to find a dead cub, less than a week old, on the spoil heap outside a main sett. There is seldom any sign of injury, although I have seen several with small punctures in the skin, probably made by the sow's teeth when the cub was carried out. It is possible that these cubs had died through being lain on. R.S. Elliott wrote to me about one cub, about the size of a small rabbit, which he found dead outside the sett; there was a mass of bedding around the entrance and it appeared to him that the cub had been cleared out with the bedding.

Cubs may be taken by wolves, lynxes and wolverines in regions where they occur, and eagle owls are said to take cubs occasionally. Golden eagles may also take them, but these are likely to be as carrion. However, it cannot be ruled out that diurnal raptors such as eagles and buzzards may very occasionally kill cubs in the late evening.

Starvation

Starvation is an important cause of cub death. It occurs mainly in the period following weaning particularly in extremely dry summers such as 1975 and 1984 when mortality was marked. I suspect that starvation is a much more common cause of death than is generally realised. Eric Ashby considers it quite usual in the New Forest area, especially in parts well away from farmland where there is a relative lack of earthworms in the acid soils.

Cheeseman found that in Gloucestershire in August 1984, following the severe drought, that most of the 70 badgers caught for sampling and release were in poor condition with low average weights. He also found animals which had died of starvation. In contrast, by November 1984, following an autumn of food abundance most were in good condition with higher average weights than normal for the time of year.

Disease

There are no records of major disease epidemics wiping out large numbers of the population; indeed it seems unlikely that disease plays an important part in population control.

In Britain, the most important disease afflicting badgers is tuberculosis due to the bovine strain of tubercle bacillus, *Micobacterium bovis*. This was first diagnosed in 1971 for a badger found dead on a farm in Gloucestershire where several cases of tuberculosis had been diagnosed in cattle. Over the next few years evidence accumulated that badgers in some areas were acting as a reservoir of tuberculosis. The implications of this will be discussed in Chapter 12.

Diseased badgers at post mortem showed a variety of organs affected (particularly in the later stages of the disease), but the lungs and kidneys were the most usual and some had become infected through bite wounds inflicted by other diseased badgers. In advanced stages of the disease derangement of behaviour has been observed. Sick badgers have been found lying up in pigsties and other farm buildings and been seen foraging during the day near houses and farms, apparently searching for any easy source of food. One animal when approached by a farmer was seen to stumble and bump into tufts of grass and branches. On post mortem severe tuberculosis brain lesions were found. But such cases are rare.

In an analysis of causes of death of badgers from Gloucestershire and North Avon, Gallagher and Nelson (1979) found that the great majority were due to road accidents, others had died of tuberculosis, non-tuberculosis bite wounds, starvation and arteriosclerosis.

Cheeseman found that for his research area in Gloucestershire for the years 1979-83 of the 45 marked badgers found dead or killed, 23 had died of road accidents, five from tuberculosis, four from starvation, five had been killed and eight died from unknown causes. There were 22 males and 23 females in the sample of which seven males and three females proved to be infected by the tubercle.

Pitt (1941) has recorded cases of pharyngitis/tonsilitis, and an acute respiratory condition of unknown origin was observed by Humphries in a colony of seven badgers near Cheltenham. To quote from his account:

On the 18th April shortly after the arrival of a strange badger all the badgers appeared healthy, but the next night coughing and wheezing was heard before they emerged. On the 20th a sow emerged very late, wheezing loudly and attempted to reach the dung-pit which was on the bank above the hole. She was unable to climb far up the bank and slipped back, rolling over. After a short rest she dragged herself slowly down the hole. Several other badgers were suffering less severely on this date. By the 22nd all the badgers except the stranger had the disease, but the sow had almost recovered and was able to emerge as usual. No trace of the illness remained on the 25th.

In Europe, badgers are known to contract rabies, although the chief carrier is the fox. However, in the serious epidemic of rabies in Germany (1950-3) over 100 badgers were affected (Zunker 1954). They are also susceptible to anthrax.

Osteomyelitis could also be a cause of death, especially in old animals. I have examined several skulls where there was gross malformation of bone associated with the socket of a broken canine which appeared to have been the primary locus of infection.

Parasites

Parasitic lung worm infestation has been reported from Germany (Schlegel 1933) and the disease proved fatal in a number of cases. No such condition has been recognised in this country. A number of species of gut parasites

Figure 11.2 Three common badger ectoparasites. 1. Louse, Trichodectes melis ♀ *(ventral); 2. flea,* Paraceras melis*; 3. tick,* Ixodes hexagonus ♀ *(dorsal). Actual sizes are indicated alongside*

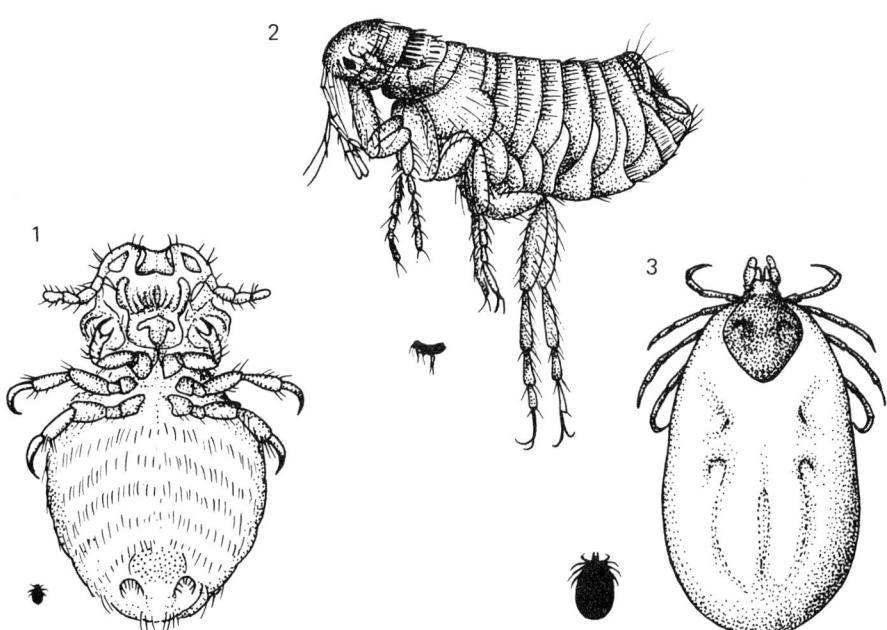

have been recorded, in particular, nematodes and tapeworms. These are unlikely to cause trouble to the animals. It is not surprising that one of the tapeworms has the earthworm as its secondary host and another is believed to be acquired through eating dor-beetles.

Ectoparasites can be a problem and might account for the scratching behaviour so commonly observed in badgers. Bedding harbours many of these parasites and its periodic removal and replacement with fresh material probably assists in keeping down the population of ectoparasites.

Fleas, lice, ticks and mites have all been found on badgers, the commonest being the biting louse *Trichodectes melis* which is specific to the badger. In emaciated or debilitated animals, the louse burden and indeed the general ectoparasite infestation is often quite enormous. The skin may sometimes appear to be seething with the small white lice and vast numbers of eggs, which are attached to the hairs, may be seen.

The flea most commonly found is *Paraceras melis*, but mole and hedgehog fleas also occur which appear to be accidental vagrants transmitted to the badger when feeding on the more typical hosts of these species. Two other species of flea occur, but only rarely: *Chaetopsylla trichosa* (a single record from Scotland, but more Continental badgers) and the human flea, *Pulex irritans* which occurs on the fox. Perhaps the few records of this flea on the badger may be accounted for through badgers cohabiting with foxes.

Ticks are less commonly found on badgers, although prevalence varies with district. Hancox (1973) found ticks on 40 per cent of the corpses he examined near Oxford, but from my experience of Somerset badgers, I consider this an unusually high incidence. Ticks are most often found on the ears, above the tail and on the inside of the thighs. Three species occur fairly regularly, the dog tick, *Ixodes canisuga*, the hedgehog tick, *Ixodes hexagonus* and the sheep tick, *Ixodes ricinus*. There are also records for *Ixodes reduvius* and *Ixodes melicola*. I was told by Gordon Thompson that *Ixodes canisuga* was originally assumed to be a parasite of sheep dogs in the north of England and in Scotland, but he considers the true hosts are foxes and badgers and that the parasitising of dogs was an adaptation, the kennel being comparable with the sett or earth.

Mark Fisher, writing to me about the badgers near Coniston in Cumbria, said that in his opinion one of the reasons why badgers did so well there was the absence of sheep ticks which were rife in some other localities in the county. He cited the case of two cubs found in the open in Nab Scar, Grasmere, wrapped in sheep's wool and covered in ticks. The cubs were so weak that there was no hope of their survival. Another instance was described by F.S. Port in a letter to *The Field*. He found a dead full grown sow with 'the whole of the face, neck, head and shoulders covered in sheep ticks—up to 1,000 of them'. It is not possible to say whether they were the cause of death, but if not they were surely a contributory factor.

Mites occasionally infest badgers. John Sankey told me how one of his badgers after being at liberty for five weeks, returned with a local infection of mites. He was unable to remove them and a bald patch developed in the middle of the back, but two applications of benzine hexachloride dispersed the infestation and the hair grew again in two months. I have also known of a mite infection of the ear. This causes intense irritation and may give rise to torn ears as a result of much scratching.

There have also been several reports of badgers suffering from mange, although this is not nearly so usual in badgers as in foxes. A. MacFarlane, who was head keeper of an estate in the north of Scotland, wrote to me about a badger which was lying in a fox's den and had no hair except a small tuft on its tail. A fox had used the den three years before and had also had mange. Other instances have been reported from the Netherlands.

Direct Action by Man

All the factors mentioned so far are possible causes of death in badgers although some obviously play a much greater part than others in keeping numbers down, but by far the most potent factors in population control are concerned with human activities.

For at least the past 200 years numbers have been kept down by direct human action. This has taken such forms as badger digging, hunting, trapping, snaring, shooting and gassing. During the 1960s and early 1970s badger digging increased and in some areas became a popular pastime for city dwellers who went out at weekends into the countryside for the purpose. This development coincided with an increase in popularity of such breeds of terrier as the Jack Russell (Murray 1968). In consequence many setts were destroyed and the badgers killed.

Since the abolition of the gin trap, the passing of the Badgers Act (1973) and its amendment in the Wildlife and Countryside Act (1981) some of these direct pressures have been reduced. Badgers are now fully protected by law (see Appendix III). However, enforcement is difficult and much persecution still takes place. Although badger digging is illegal it is still widely practised in spite of a number of convictions resulting in heavy fines. Badger protection groups have been set up in some areas to combat this menace.

Badger hunting with dogs, above ground after the badgers had left their setts to forage, has largely ceased, but 'lamping' late at night using spotlights from vehicles is a growing menace. The lights locate and dazzle the animals and guns, nets and dogs are used to kill or capture them.

Gassing with cyanide preparations such as 'cymag' has been widely used in spite of its being illegal to kill badgers in this way. It was used as a routine measure on certain farms and by some shooting syndicates for killing foxes and badgers. The loophole in the law is that it is legal to kill foxes by gassing, and as foxes often inhabit badger setts it can be pleaded that by gassing these, the intent was to kill the fox. It is not possible to assess how many badgers are killed by this means today, but it must still be great.

It is ironical that with the abolition of the gin trap, the practice of snaring badgers has greatly increased. Badgers are caught very easily by this method and I know of one wood in the Quantock Hills where a thriving colony of badgers was totally snared out by a farmer who, through ignorance or prejudice, considered they were a potential danger to his lambs. The action was entirely unnecessary.

Snaring often results in much cruelty, as badgers are very strong and may go off with the snare round them to die a lingering death. I have several times seen badgers which have died in this way with the snare deeply buried in their flesh. Many badgers, of course, are caught in snares set for foxes, but many snares are deliberately set for badgers and large numbers are killed: this again is illegal, but is practised commonly.

So although legislation and public opinion has done much to reduce pressure on the badger, direct action against them still remains a very significant cause of death and a threat to their survival in some areas.

Traffic

Many badgers are killed on the railways where the track separates one sett from another and ancestral paths connect them. Charles Eyre, a signalman who worked near Bath in Avon told me that an average of four badgers were killed on the line every year less than 50 m from the signal box where he worked. There was also another crossing place in Somerset, near Taunton, where badgers were killed regularly by trains. These instances are typical of many parts of the country and in total must account for a large number of badgers. But, the most serious threat is ground-level electrification. During the few months following the electrification of the line between Ashford and Deal in Kent, Michael Alcock told me that 68 badgers were electrocuted, and over a longer period of time as many as 200 died in this way. Similarly, in Hampshire, Arthur Jollands picked up eleven badgers on a 5 km stretch of line near Alton in a single year and six more in the first three months of the next. Also, when the line was electrified in the New Forest area, 30 badgers were killed during the subsequent year over a 14.5 km stretch.

The number of deaths caused in this way do become fewer after a few years, but whether this is due to learning or reduction in numbers in the neighbourhood is difficult to know. Learning can only come through experience of the death of another badger, as contact with the rail is always fatal. A most intriguing aspect of the problem came to my notice through G.P. Knowles, who knew a retired army officer who lived by the railway at Camberley in Surrey. Before electrification came to that line, foxes had an earth in the embankment and there was a badger sett just inside the wood nearby. Both species used to wander across the line regularly and when electrification came, there were many casualties. However, some time after this event, his informant saw a badger crossing the line in a zig-zag manner without getting harmed. Later he saw a fox do likewise, and subsequently many of both species crossed in this way. On inspection of the line he found that they passed between the gaps in the live rails! One wonders if these animals are capable of detecting the magnetic field surrounding the rail? Unlikely perhaps, but not impossible, as some animals can orientate their movements in relation to magnetic fields.

A very large number of deaths occur on the roads of Britain each year. How many is impossible to assess, but an indication of the scale of mortality may be gauged by the fact that following a public appeal in connection with the tuberculosis investigation, 800 road casualty carcasses were sent in during a single year from south-west England. The numbers killed must have been much higher than this as presumably only a proportion were reported. Earlier on during the investigation another sample of 442 road casualties provided valuable information on badger activity throughout the year. I am grateful to the Ministry of Agriculture and in particular to John Gallagher for allowing me to use these data.

In this sample, the number of male and female casualties was almost identical (220 males: 222 females), but there were definite seasonal differences. In the periods February–April and August–September, more males were killed, but for May–July, more females (Figure 11.3).

The greatest number of casualties, irrespective of sex, occurred between February and April—38 per cent (22 per cent males and 16 per cent females). This is a period of great activity, especially among the males, as this is the main mating season, when they spend more time patrolling their territories. It is also the time when young males are reaching maturity and may be seeking out territories of their own. Furthermore, it also coincides with the breeding season and if severe weather occurs, the sows too may have to move further than usual to find enough food. Sow casualties at this season are of particular significance as the majority are lactating and their litters would also die. In 1969, of 13 sows I received during this period, eight were lactating and the others were yearlings.

There is another rise in casualty figures in August and September. This is a time when feeding is of primary importance, territorial limits are less strictly observed and rather more animals are living rough away from their main setts for varying periods.

If the casualty graphs for each of the four years 1973–6 are compared, those for 1973, 1974 and 1976 follow the same general pattern with peaks in February–March and August–September, but for 1975 the picture is quite different, due to a rise in May and June, leading to a spectacular peak in July. It was during this period that the severe drought affected the badgers so greatly and altered their usual behaviour pattern. So the abnormal rise in casualties at this time was probably due to greater activity related to the quest for food. Because 1975 was so exceptional, I considered it would be misleading to include these data in Figure 11.3, but they have been included as an inset for comparison. The drought of 1976 had far less impact on the badgers as its effects became apparent much later than in the previous year and by then cereals and fruits such as blackberries had ripened.

Poisons

There are records of badgers having died as a result of eating poisoned bait, probably intended for other animals. These include strychnine and the

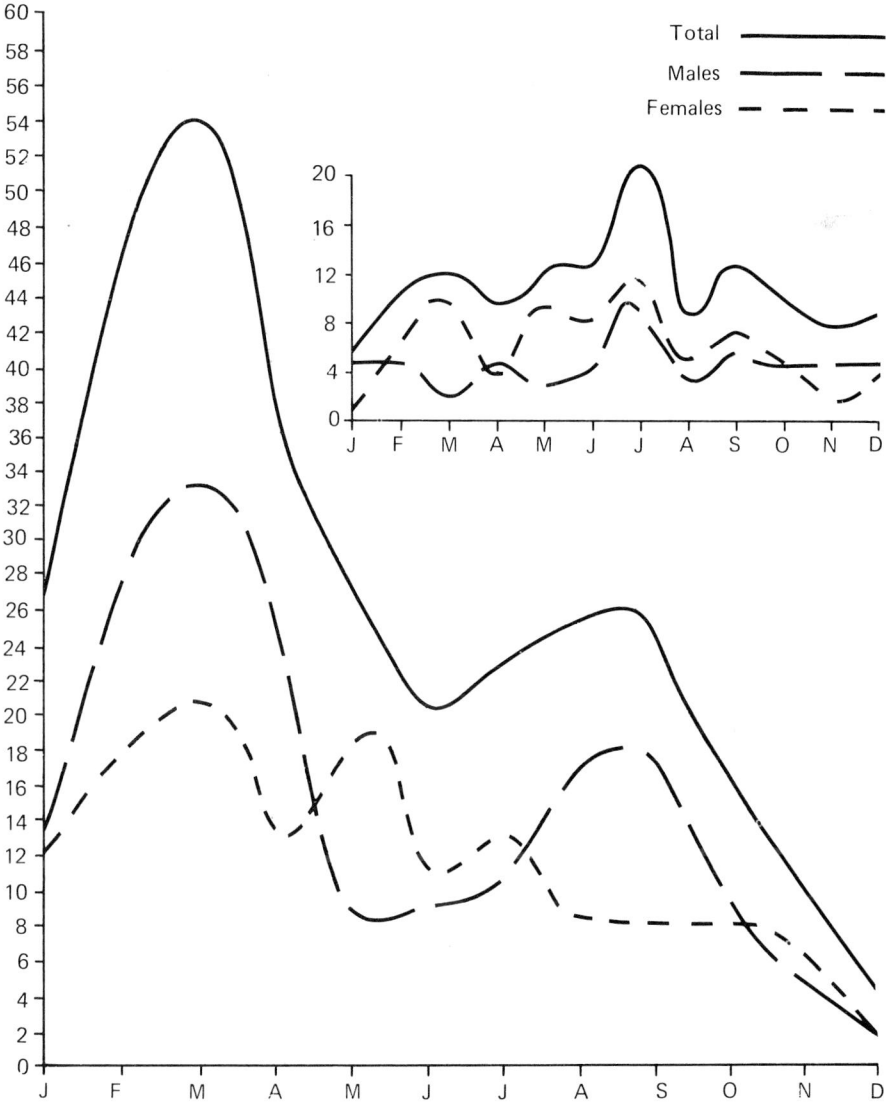

Figure 11.3 Badger road casualties (316 animals), 1973, 1974 and 1976 examined at Gloucester Veterinary Centre. Inset: 1975 (126 animals) showing presumed effect of the severe May–July drought (see p. 192). *By courtesy of Dr J. Gallagher*

rodent poison warfarin. However, death by such poisons seems to be rare.

The effect of various pesticides on badgers has been extremely difficult to assess, and no firm conclusions may be drawn except in a few well-documented instances. Great anxiety was expressed among badger watchers because there appeared to be a considerable drop in the number of cubs produced in some parts of Britain during the period 1962–9. This was a time when the persistent organo-chloride pesticides were being used in considerable quantities.

Where such pesticides as DDT and dieldrin were used regularly, large residues of these substances were accumulated in the tissues of earthworms, although they themselves appeared to be unharmed by them. So, it was argued that as badgers ate such large quantities of earthworms, the effect would be passed on. To investigate this hypothesis, a number of badger livers and fat samples were analysed, but the majority of them showed only trace residues, certainly not enough to cause death.

However, D.J. Jefferies (1968) found that between 1964 and 1968, of the 17 badger carcasses that he received for analysis at Monks Wood Experimental Station, at least six had certainly died from dieldrin poisoning and circumstantial evidence suggested that another six had died from the same cause. This, of course, was not a random sample of badgers found dead, but it does indicate the dangers of such pesticides. It was concluded that the probable chain of events was that seed dressed with dieldrin had been eaten by pigeons which had died as a result and been eaten in their turn by badgers.

The symptoms of dieldrin poisoning in the badger are apparent blindness, lack of appetite and convulsions, especially of the jaws. Badgers suffering in this way were seen above ground in daylight; they took little notice of people and sometimes wandered in circles, bumping into objects. Incidentally, foxes were affected by dieldrin poisoning much more than badgers, 1,300 dying in the winter of 1959–60 in the eastern counties of England alone.

Whether such pesticides absorbed in sub-lethal quantities interfered with the reproductive process in badgers (as it is known to have done in such birds as herons and buzzards) is now impossible to prove, but this could have been the cause for a decline in cub numbers in the late 1960s.

Fortunately, the total ban on this type of persistent pesticide in Great Britain has removed this particular hazard.

12 Badgers and man

Badgers, in common with most wildlife today, are greatly influenced by human pressures, either directly or indirectly, and their survival in many regions is no longer mainly dependent upon their ability to adapt to changing conditions, but is largely determined by man's attitude towards them.

Inevitably, on a crowded island like Britain, any wild mammal species, especially one of fairly large size, conflicts in some ways with man's interests. The badger is no exception. Attitudes vary as to what action should be taken when these conflicts arise, but today it is increasingly realised that wildlife is a priceless asset which should be conserved, even if this entails slight economic loss or inconvenience. The majority of farmers and landowners take this view and are tolerant of the presence of badgers on their land and many welcome them and guard their interests jealously. This attitude is justified, as scientific evidence makes it clear that the activities of badgers are mainly neutral to man's interests; in some districts they may do a little harm, but they also do good by reducing the numbers of pests. It is all a matter of balance. Under normal conditions badgers need no control, but exceptionally, if the population density becomes too high, they may become a matter for concern and reduction in numbers may be justified. The Badgers Act (Appendix III) clarifies the steps which may and may not be taken if badgers have to be controlled.

Conservation of any species involves maintaining the right balance and this can only be achieved through an enlightened attitude of understanding and tolerance. Extreme sentimentality on the one hand, and entrenched prejudice on the other, do nothing to achieve a sensible balance of interests. In this chapter, I would like to face realistically the points of conflict which may occasionally arise between badgers and man and emphasise the positive steps that may be taken for the benefit of both parties. But first let us look at the urban situation.

URBAN BADGERS

Badgers are remarkably adaptable animals. Not only have they succeeded in colonising a great variety of rural habitats, but they have also clung on with extraordinary tenacity to many urban situations and become well adapted to urban life.

BADGERS AND MAN

In 1969, W.G. Teagle, in his admirable survey of badger distribution within a 32 km radius of London, recorded 164 setts he knew personally and at least another 110 which were reported on good authority. Some of these setts, which are still active today, are situated in commons and parks within a remarkably short distance of Central London. It is doubtful whether they still occur in the Hampstead area, although reports of sightings have been published from time to time, certainly up to 1967. The most memorable sighting on Hampstead Heath occurred during World War I when a pilot made a forced landing there in the early hours and came face to face with a badger!

There is one main road in London which badgers often cross, once the stream of buses and cars has subsided, in order to get from their sett in the grounds of a teacher's training college to the gardens of a convent—where on one occasion they drew attention to themselves by attempting to undermine a statue!

Many other cities such as Bristol, Bath and Birmingham have their quota of setts and in the early morning badgers are commonly seen by police patrols and late-night travellers. Stephen Harris (1982, 1984) made a splendidly comprehensive study of the Bristol badgers. He found 80 setts in a 5.5 km² arc of the north-western part of the city. Forty were in strips of woodland, scrub or bramble-covered banks, 29 in private or semi-public gardens and the remainder in places such as horse paddocks and under outbuildings.

Compared with rural areas he found that home ranges overlapped more and territorial boundaries were not well defined; it was the area around a sett that was heavily marked by dung pits rather than the perimeter.

Earthworms, although readily available on lawns, playing fields and open commons, were exploited far less than in rural areas, the badgers relying on a greater range of foods. Certain categories predominated according to season. Thus from January–May this was scavenged food (24 per cent over the year), June–July, invertebrates other than earthworms (20 per cent), August–November, fruits and vegetables (35 per cent) and December, earthworms (18 per cent). Scavenged items included food taken from dustbins, compost heaps, bird tables and food specially put out by householders for the badgers.

Compared with rural areas emergence is characteristically later, usually well after dark; they play very little, go straight off after emergence and restrict bedding collection mainly to the autumn and early spring when nights are longer. The major cause of death is from traffic.

Occasionally urban badgers have got into difficulties. One was found trying to get into the cellar of a girls' school and another was found in the crypt of one of the churches. Stephen Harris tells the story of a badger which lived under the floorboards of an occupied house, entering through a broken airbrick. When several floorboards had been removed to lay new piping it would come into the rooms to eat the cat's food. One night the lady of the house was awoken 'to absolute bedlam: the badger was chasing the cat around her bed!'

Some years ago, patients in sanatoria in both Surrey and Hampshire had their lives greatly enriched by nightly visits from badgers which would come right up to the windows of their wards. One patient was able to coax a badger to within 1.5 m of his bed by laying a trail of scraps.

However, when badgers are in such close proximity to man, they may sometimes cause annoyance by digging shallow pits in the lawns after earthworms and cockchafer grubs, digging up flower bulbs or carrots or eating soft fruit. If this happens they may be kept away by soaking a rope in some repellent such as old diesel oil or renardine and supporting it on short sticks about 125 mm high across the line of entry into the garden. The rope should be re-soaked every few weeks, especially after rain. This technique is much more effective if it is used before entry becomes a regular habit; for example, before the fruit ripens. In more serious instances an electric fence, as for cattle, but set at 125 mm from the ground, works well. Electrified flexinet has also been proved extremely effective. Sheep netting erected as described on p. 199 will also keep badgers out.

THE PROBLEM OF TRAFFIC

As already discussed (p. 191) road and rail traffic account for the deaths of a very large number of badgers and this is particularly true of main roads and motorways which cut through good badger country.

Michael Clark, who prepared the artwork for this book so attractively, has been much concerned with the conservation of badgers in Hertfordshire. He suggested (1970) that because badgers used regular routes, some losses could be avoided by diverting the badgers into pipes underneath the roads. The Highways Department in Hertfordshire considered his detailed plans and general agreement on procedure was reached (see Appendix II for details).

The first project to be completed in Hertfordshire involved the use of an Armco stream culvert which carried a steady flow in times of heavy rainfall. A concrete ledge was constructed along one wall of this culvert for the badgers to use when the water level was high. The fencing on either side of

Figure 12.1 Many badgers are killed crossing roads

the culvert was fitted with galvanised netting as this not only prevented the badgers from crossing the road, but guided them towards the culvert. However, it was found that where main badger paths had come to where the fence was erected, the badgers would dig underneath unless the netting was turned outwards at soil level and covered. Materials were paid for by the road construction unit, but all the work was done by volunteers including members of a Workers' Education Association group, Conservation Corps enthusiasts and local badger watchers. The project was a considerable success and badgers now use the ledge regularly (Figure 12.2).

Following Clark's original paper, Jane Ratcliffe with the help of her husband (who is a civil engineer), put forward plans for a rather similar structure under the M53 motorway in Cheshire. This too used a culvert, towards which the badgers were directed by netting fastened to the wooden post-and-rail fencing. Specifications for this construction have been published and caused much interest (Ratcliffe 1974).

Tom Wilton, a civil engineer concerned with the construction of the M5 motorway in Somerset, was also very keen to reduce the destruction of wildlife on the roads. He approached me in 1973 about the possibility of putting in badger tunnels where these were appropriate. We surveyed the proposed route of a section of the M5 together, mapping out the position of badger setts and the main paths in relation to them. The most vulnerable places are undoubtedly where two setts are close to the road, but on different sides of it, or where the sett is on one side and a main feeding ground on the other. One such place was found west of Wellington in Somerset and as the

Figure 12.2 Stream culvert of the Armco galvanised steel type under a by-pass in Hertfordshire, showing raised badger path (left). *Water level nearly reaches the path when in flood*

Figure 12.3 Badger-proof fencing. a) Timber post-and-rail fence with wire dug in. Badgers still dig under this; b) wire laid flat on surface of ground, lightly turfed. Badgers may dig at base, but this reinforces rather than weakens the strength of the wire, as long as the gauge is strong enough

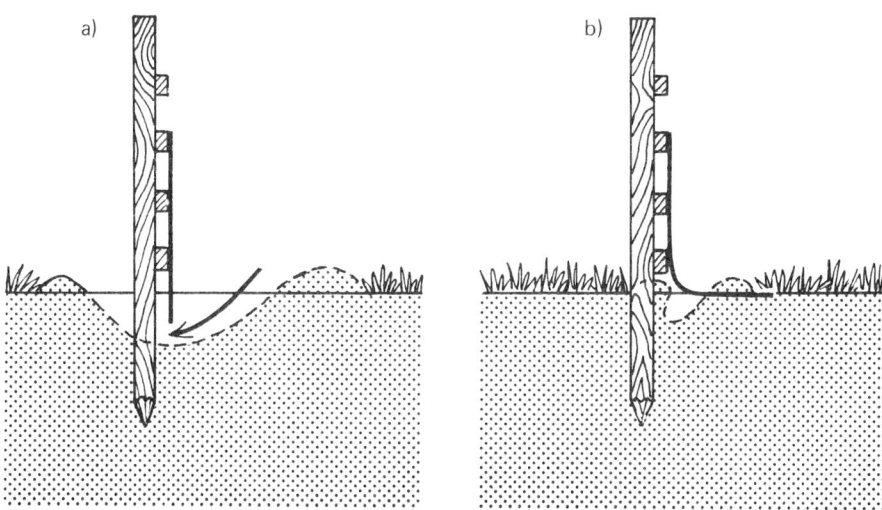

motorway was to be on an embankment at that point, it was a suitable place for a tunnel. The plans were passed, as it was accepted that this was not only for the benefit of the badgers, but that an adult badger knocked down by a light car travelling at 110 km/hour was also a major hazard to the motorist.

Figure 12.4 Badger tunnel built under the M5 motorway in Somerset. The netting on the post-and-rail fence helps to guide the badgers to the entrance and stops them from crossing elsewhere

Unfortunately, there were long delays before work on this section of the motorway was started, but by the summer of 1976 the tunnel was in position at a place where a main path from the sett met the road. At this point a concrete drainage ditch ran parallel to the motorway, so wooden planking had to be erected to act as a bridge over which the badgers could pass. The tunnel consisted of a concrete pipe 600 mm in diameter and 50 m long. Sheep netting was fastened to the fencing for some distance on either side to help guide the badgers towards the entrance. At first the badgers were reluctant to use it so an aniseed scent trail was laid up to and through the tunnel to encourage them. But it was not until numbers increased in the sett nearby that they started to use it regularly. Finally they claimed it as part of their territory by making dung pits near both ends of the tunnel.

In the new town of Milton Keynes in Buckinghamshire, elaborate plans for underpasses were worked out by the field ecologist John Kelcey, as there were a number of badger setts within the complex. Enthusiastic members of the local natural history society helped considerably over the project, as they did in Hertfordshire, where the results would not have been possible without the combined efforts of many people.

In some situations it is better to attempt to stop badgers crossing the road at all. This involves long lengths of appropriate netting fixed to post-and-rail fencing as specified in Appendix II.

A special problem occurs when the route of a proposed road passes directly through a badger sett. The best method is to stink the badgers out, so that they move of their own accord to an alternative sett within their home range. Unfortunately, in some localities no alternative setts exist, but in most, this method can be used. To do this it is best to wait until the time for bulldozing is imminent. The sett should then be visited when all the occupants are expected to have left, between 23.00 and midnight. Every hole is then stuffed with newspaper soaked in old diesel oil or other repellent and then lightly stopped with earth. Later that night when the badgers return they will detect the smell and not attempt to re-open the holes, but go to their alternative sett. This method should not be used between February and May when young cubs may be below.

BADGERS AND FARMING

Under most circumstances, there is no conflict between the interests of farmers and badgers; the little harm they do is often more than balanced by their destruction of pests. However, in a few places where population density has become unusually high or when weather conditions are exceptional, badgers may find it extremely difficult to get enough food and may cause some damage. To some extent this depends upon the type of farming.

The poultry farmer normally has little to fear from badgers, although poultry killing certainly occurs under exceptional circumstances (p. 128). Large-scale poultry farms are seldom affected as most losses are suffered when poultry are inadequately housed. Entry to the henhouse is usually effected by pushing the snout under the drop-hatch which can easily be prevented by letting the hatch slip into a groove at the bottom of the door.

It is bad practice to throw out dead hens on to a refuse heap near a henhouse, as this may lead to trouble. If badgers find this source of food, they may acquire a taste for poultry and so cause them to attempt to get at the living birds nearby.

For the arable farmer, the chief problem is when cereal crops are flattened and grain consumed. This may happen during the few weeks prior to harvesting oats and wheat. The matter has been discussed fully in a previous chapter (p. 132). However, it is worth emphasising here that most damage occurs if a cereal crop is grown near to a wood or copse in which there is a sett. If the farmer has a choice, and grows barley in such a field, the badgers will usually avoid this and the crops of oats or wheat further away will receive far less attention.

Occasionally, badgers will open up holes in fields which connect underground to setts in hedgerows or copses. This can be a hazard, especially with heavy machinery, as the roof of the connecting tunnel may collapse and the tractor sink into the ground. This problem is best tackled as soon as it is noticed, as it is far more difficult to remedy once the field entrances are well established. Again, the best way is to try to stink the animals out from that particular part of the sett by stuffing up the entrance with newspaper soaked in repellent and firmly stopping the hole with earth.

BADGERS AND BOVINE TUBERCULOSIS

For the dairy farmer badgers are no worry at all under normal circumstances, the only exception being where bovine tuberculosis has become endemic in the badger population.

This problem needs to be considered in relation to the history of this disease over the past forty years. In 1946 it was reliably estimated that 18 per cent of all cattle in Britain and 30–35 per cent of all cows were infected. Following the Ministry of Agriculture's campaign of slaughtering all cattle which reacted to the tuberculin test, the disease today in Britain is very low indeed (around 0.02 per cent). However in south-west England, after a steady decline comparable with other parts of the country, the incidence has remained for the past twenty years 5–10 times as much as for the remainder of Britain (0.04 per cent compared with 0.006 per cent in 1983).

This difference suggested that there was some unusual factor responsible and a wildlife reservoir of infection was suspected. Suspicion fell on the badger when the causal organism, *Micobacterium bovis* was isolated from a badger carcass in 1971. Evidence then began to accumulate that badgers in some areas were acting as a reservoir for tuberculosis and that under some circumstances it could be passed on to cattle. In the light of this evidence and under strong pressure from the National Farmers Union, the Ministry embarked on a programme limited to those areas where cattle reactors had been found and where the disease had also been diagnosed in badgers. The aim was to eliminate bovine tuberculosis from the badger population in order to remove the hazard to cattle and possibly humans. Hydrogen cyanide gas (as previously used for rabbits), force-pumped into the sett as a powder, was used for the purpose as at that time it was considered the most humane and effective method of killing badgers underground. The programme started during the winter of 1975/6 under licence from the Ministry and was carried out by specially trained teams taken on by the Ministry and under the supervision of Ministry officials.

Understandably, such drastic action against the badger was looked upon with horror and dismay by naturalists and conservationists, but it was recognised by all responsible conservation bodies that a serious situation had arisen and with regret, the Ministry's action was considered justified, not least because individuals might take the law into their own hands. Nevertheless, much concern remained that the method might prove ineffective and become open-ended. This anxiety was recognised during the relevant debate in the House of Lords, so a non-statutory consultative panel was appointed to keep under review the general question of tuberculosis in badgers and its relationship to the disease in cattle..

The gassing programme continued until June 1982 except for a moratorium between September 1979 and October 1980 when Lord Zuckerman was preparing his report. Experiments to discover what concentrations of gas were necessary to kill badgers effectively cast doubts on the humaneness of the methods employed and the Minister for Agriculture, the Rt Hon. Peter Walker, MP decided to abandon the method. Cage trapping followed by shooting was generally agreed to be the most effective and humane alternative and this was put into practice. The great advantage of this method over gassing was that the carcasses became available for research and the proportion and distribution of diseased animals could be determined.

No other wildlife reservoir of the disease has been discovered although up to December 1984, 4,187 animals of 21 species of mammal had been analysed. Of these, *M. bovis* was recovered from two deer (out of 65), five foxes (751), one mink (172), two moles (161) and five rats (409). It was considered that in the infected foxes, rats and moles the disease was not progressive and hence non-infective. Domestic animals such as dogs, cats and pigs are possible carriers, but there is no evidence that these are a source of infection.

Let us now consider the disease in badgers. At present, there is no effective method of detecting *M. bovis* infection accurately in live badgers. Clinical examination together with bacteriological examination of feces, urine, sputum and bite wound samples is the best available approach, but it is time consuming, not always accurate and will only identify infectious (not infected) animals.

Animals can remain infectious for some time. In the laboratory one badger survived for 3.5 years and five for 1–2 years whilst excreting *M. bovis* bacilli (Little *et al* 1982). Monitoring disease progression in a natural population, Cheeseman *et al* (1981) found that three badgers were infectious for between one and two years.

The course of infection in badgers is very variable and seems in part to depend on the route of transmission. Bite wound induced infections tend to progress rapidly, whilst infections caused by inhalation seem to progress at a slower rate (Gallagher and Nelson 1979). Excretions swallowed from the respiratory system enter the digestive tract. Neither complete recovery from infection nor the acquisition of immunity appear to occur, although the infection may become latent (inactive) for periods during which time the bacilli are not excreted in detectable quantities. There are also badger carriers, apparently healthy, infected, but not infectious animals.

The most common route of transmission is by inhaling live bacilli. When a sow is infectious, the close contact with her cubs places them at high risk of infection during early life. Badgers in general may contact the infection directly from other badgers or indirectly from contaminated pasture. Diseased animals can contaminate the environment heavily with bacilli passed via feces, urine, sputum and suppurating bite wounds. Where there are kidney lesions, up to 300,000 bacilli per ml of urine can be passed. Transmission can pass from badger to badger by biting: for example in a survey in

Gloucestershire, 14 per cent of the observed cases of disease in badgers were thought to have arisen from bite wounds (Gallagher and Nelson 1979). Animals with severe lung lesions have high concentrations of bacilli in their saliva and sputum and are therefore highly likely to transmit infection by biting susceptible badgers.

The social behaviour of badgers influences the effectiveness of disease transmission. Warm, dark, moist environments, such as those found in setts, appear optimal for bacterial survival, as does communal living for disease transmission. As social animals relying on scent for identification, badgers frequently nuzzle and examine each other. Badgers forage on pasture for earthworms which are their main source of food and the high water content of the worms results in frequent badger urination. Frequent visits are made to dung pits on territorial boundaries to defecate. All these activities facilitate the spread of the disease within social groups.

Disease transmission between social groups can occur through dispersal, particularly of yearling males; or from the extended wanderings through neighbouring territories of severely diseased badgers (Cheeseman and Mallinson 1979). Aggressive behaviour, particularly by males, increases during the breeding season as part of territorial defence and at that time adult males sometimes move into adjoining territories seeking sows in oestrus (Cheeseman and Mallinson 1979). There is little information to indicate whether there are seasonal changes in transmission, but spring appears to be the time of greatest activity.

The result of these effects is that bovine tuberculosis appears to be stable and able to persist in high-, moderate- and low-density populations. The disease is probably endemic in many badger populations throughout Britain, but is particularly prevalent in areas of good badger habitat in south-west England; the proportion of badgers having the disease in a high-density population being likely to be higher than in a low one.

Now let us consider aspects of transmission of the disease from badgers to cattle. Current evidence suggests that badgers play a significant role in this process. There is a correlation between the location of herd breakdowns and tuberculous badgers nearby in a majority of cases. Also under experimental conditions badgers from an infected population kept in a covered yard along with calves were proved to transmit the disease both to other badgers and to the calves, but only after 4–5 months. There is also convincing evidence concerning two large areas with particularly bad histories of breakdowns and high badger infection at a farm in Dorset (12 km^2) and Thornbury (100 km^2). Here, badgers were totally eradicated and since then no breakdowns have occurred—a period of 3–6 years at the time of writing.

Between 1971 and 1983 the average level of prevalence of the disease in badgers for the whole of the south-west and mainly from the vicinity of breakdowns was 14 per cent, while in a large sample submitted by the public from the same area (mainly from road accidents and hence a more random sample) it was 3.4 per cent for the same period. Figures from Gloucestershire, Wiltshire and Avon combined, between 1971 and 1976, showed a 20 per cent prevalence and between 1977 and 1982 after persistent control operations, it was 10 per cent; a reduction in the rate of infection of only 50 per cent. However, in that area and during the same period, there was a more substantial reduction in the incidence of cattle herd infection.

Many of the factors which facilitate the transmission of the disease from badger to badger, aid badger to cattle transfer. It is also of particular importance that badgers forage for their main food item, the earthworm, on cattle pastures. It seems that more than one badger is usually needed for effective transmission of the disease to cattle. Factors such as high disease prevalence in the badger population, high cattle densities in pastures shared by badgers and the practice of certain types of farm management will facilitate transmission. Although these factors could operate in large areas such as the Cotswolds, they could also apply extremely locally wherever there is a focal point which attracts both cattle and badgers, such as the presence of cattle feeding troughs or farm buildings.

In this connection, the excellent research over a 3-year period by Paul Benham (1985) on badger/cattle interactions is of considerable importance as its application to farming practice might reduce the chance of cross infection. He found that on the whole, badgers avoided close contact with cattle; this applied to individuals, but even more so to groups of cattle. In rotational grazing systems (where cattle are restricted to a small area) badgers strongly avoided cattle areas and very rarely went within 10 m of cattle. In fields where cattle had continual access to the whole enclosure, avoidance of the cattle by badgers was more difficult and the badgers became more habituated to their presence. Under such circumstances cattle, during night grazing, would travel all over areas where badgers had recently been.

Benham found that cattle were very curious of badgers particularly if they were carrying out unusual behaviour such as collecting bedding. On being approached a badger would retreat rapidly and occasionally long chases would take place. He also observed the behaviour of cattle when badger setts were present in their pasture. Cattle spend much time rubbing their heads in mud and a sett entrance was used for this on many occasions. Such an action would cause interest by other cattle causing them to collect around the sett. To quote from his account, 'cattle heads were placed right down the hole and sniffing and licking the ground and the soil on the faces of other animals were performed'.

Work was also done to see whether badgers are attracted to eat from cattle food sources. A range of foods was made available for almost a year. It was found that badgers would avidly consume rolled barley and dairy nuts at all times of the year and proton blocks, molassine meal and maize silage occasionally, particularly during dry periods when normal food was scarce.

Benham also investigated the response of cattle to badger products. He concluded that generally cattle strongly rejected ingestion of herbage contaminated by feces and urine and this rejection was still apparent up to 28 days for the former and 12 days for the latter, although these responses varied greatly between individual cattle. A few (2–3 per cent) were unselective in their grazing behaviour and ate contaminated herbage. Most contaminated herbage was eaten when less alternative food was available.

The results from this work demonstrate that the level of contact between cattle and badgers and cattle and badger products can be manipulated by altering farm husbandry practices. A list of management recommendations designed to minimise the risk of transmission of *M. bovis* from badgers to cattle has been compiled by Paul Benham. They can be summarised as follows:

(1) Do not graze cattle hard ... as herbage becomes scarcer cattle will graze progressively nearer to contamination.

(2) In high risk areas use a rotational grazing system that contains the cattle in a small area . . . particularly in spring and autumn when badgers are most active in pasture.
(3) Prevent access of cattle to badger setts.
(4) Prevent badgers and cattle eating at the same food source . . . food troughs or stored cattle food.
(5) Exclude badgers from cattle grazing areas or other high risk sites using electric fence or flexinet.

The crucial question today no longer concerns whether badgers can infect cattle, but whether the control methods being used are effective in eradicating the disease or at least reducing it to an acceptable level. Unfortunately there is little evidence of success. What is so disturbing about the present strategy is that so many badgers are killed with so little improvement in the cattle situation to show for it. And even if, over the next few years, the disease incidence in cattle should go down significantly, it would mean that control operations would have to continue indefinitely. Could this possibly be justified?

It has been suggested that by lowering the population density of badgers sufficiently, infection in cattle could be greatly reduced. However, the model produced by Roy Anderson suggests that intensive trapping once every three or four years would be required to suppress and hold badger densities effectively at the required low levels. This would be completely unacceptable on both economic and conservation grounds.

However, an approach in addition to that studied by Benham which also holds some hope is the production of a safe, oral vaccine which could be fed to badgers in infected areas incorporated in bait. It has been shown by Drs Kahlid Mahmood and John Stanford at the Middlesex Hospital Medical School that, contrary to previous belief, a cell mediated response to the disease does occur when badgers are injected with BCG vaccine. But BCG contains live but weakened bacilli, so even if an oral vaccine of this type could be produced, permission to use it in the countryside might not be given. There is still much complex and time-consuming research to be done before this line of enquiry leads to a practical and safe technique.

For a full appraisal of the whole problem of cattle, badgers and tuberculosis the reader is referred to the extremely valuable paper by Professor Anderson and Mr Trewhella (1985) to whom I am greatly indebted and the *Wildlife Link Report* (World Wildlife Fund 1984) compiled by representatives of conservation and scientific bodies. The report accepts that badgers are a factor in the relatively high incidence of bovine tuberculosis in the south-west, but it stresses the need for more research on the epidemiology of the disease and the immunological aspects of control. Among other important recommendations it urges the Ministry to establish special 'no badger interference zones' with the agreement of farmers who, in return for ensuring no interference, would be given annual research assistance grants in addition to the 75 per cent compensation payment for all infected cattle. It is hoped that this might demonstrate whether or not, if left alone, a natural decline in the disease level in cattle and badgers would occur. This is important as it is not known whether tuberculosis in badgers is a new phenomenon or one on the way out. There is no evidence that it is spreading.

It is greatly to be welcomed that an independent, highly qualified review team are now looking into current policy and will report to the Ministry.

Their recommendations will be made public, probably before this book appears.

The problem is a serious one, but it should be kept in perspective. Fortunately it is one which only affects limited areas mainly in south-west England. On present evidence, badgers from most parts of England are in no way a hazard to cattle, and the Ministry emphasises that there is no justification whatever for farmers outside these areas of infection to take action against badgers on their land. It is illegal anyway.

BADGERS AND GAME REARING

This is a subject about which there is much prejudice. It is understandable that some gamekeepers take the line that it is better to be safe than sorry, so they kill the badgers near pheasant-rearing areas. This attitude is not justified by the evidence (the matter is discussed in detail in Chapter 8). Here it is sufficient to reiterate that the Game Conservancy considers that damage by badgers in relation to pheasant rearing is insignificant and calls for no repressive measures. Some losses do occur occasionally, but they are small.

BADGERS AND FORESTRY

The Forestry Commission recognises that the badger is the friend of the forester. Badgers certainly destroy many pests. Young rabbits are eaten in comparatively large numbers in districts where they are common, and nests of field voles and woodmice are also destroyed. In years when these rodents are abundant, large numbers are destroyed by badgers. As these mammals are the main enemies of natural regeneration, badgers are particularly helpful in this respect. Badgers also reduce the wasp menace, which is a constant irritation to foresters in autumn. The one action that is not appreciated is when a badger forces up the wire round a young plantation and lets rabbits in. This is now prevented by putting in badger gates whenever main badger paths leave a plantation. The gate was perfected by R.J. King (1964) and takes the form of a door which is free to swing both ways. It is made of heavy timber suspended from a stout wooden frame, and a wooden sill is placed below it so that rabbits do not burrow under and force an entrance. Badgers easily push through against the weight of the door,

Figure 12.5 Badger using the type of swing gate recommended by the UK Forestry Commission

but rabbits cannot enter. During erection the door is not hung in position until the badgers are using the opening regularly. The device works extremely well and the erection of badger gates is now normal forestry procedure in many parts. Details of construction are shown in Figure 12.6.

Figure 12.6 Badger gate specification. Preferably, it should be made throughout in oak. See main text for procedure

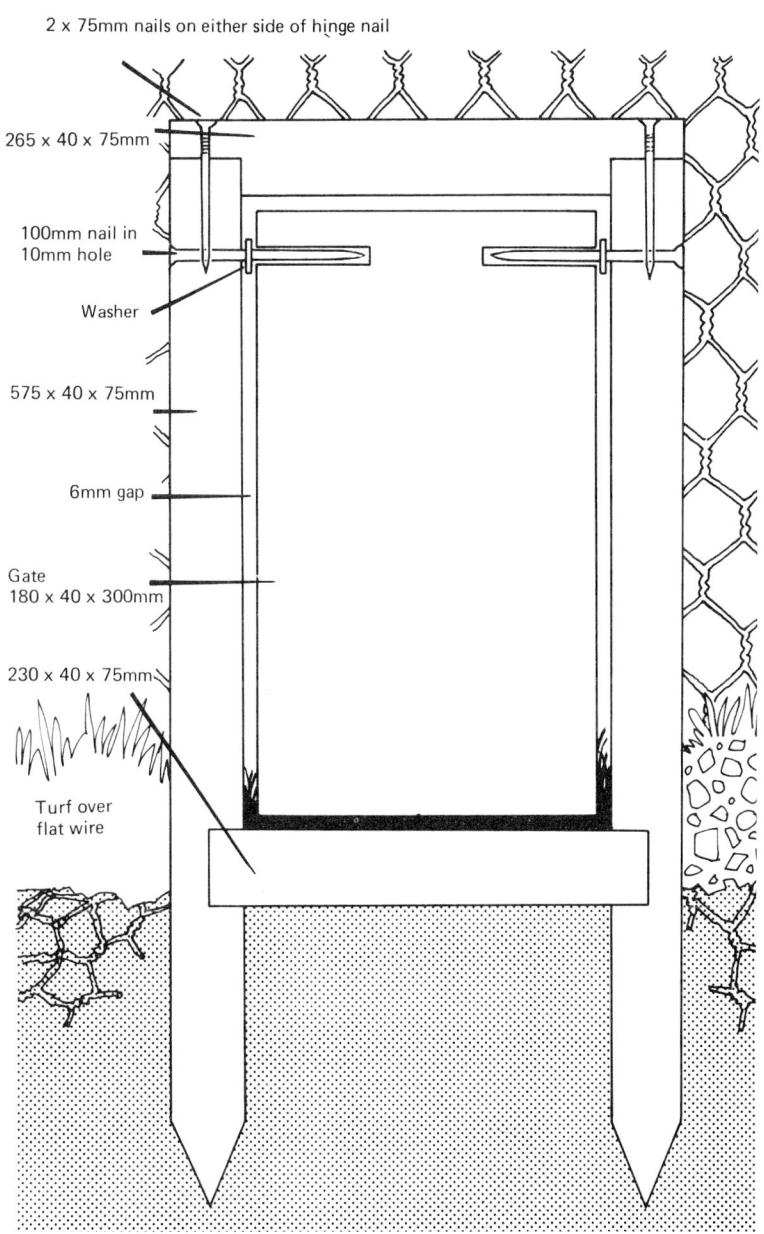

BADGERS AS A DELIGHT

Having described some of the ways in which the interests of badgers and man may occasionally come into conflict and the steps that may be taken to prevent or minimise any possible damage or annoyance that may result, it is right also to refer to a further aspect which puts us greatly in the badgers' debt. This is the considerable amount of pleasure and satisfaction that badgers give to many people.

This has been brought home to me in many ways, not least by the vast correspondence I have received about them over the past 40 years.

It is no exaggeration to say that very large numbers of people have taken up badger watching and for many it has added a new dimension to their lives. Why do people from many walks of life and of all ages spend hours, often under uncomfortable conditions, watching for that black and white face to appear at the sett entrance—often without success? The reasons for doing so are no doubt mixed and varied. For some perhaps, the very fact of being in a wood at night is a thrilling adventure, for others what appeals most is the challenge of having to pit their wits against the animal's instincts and intelligence. Curiosity is one of man's most obvious characteristics, so perhaps for some it is the desire to find out more about the lives of animals that makes them do it. For many, it may be a means of relaxation, an escape from the sophistication and materialism of our modern society, even an escape from boredom. Perhaps it also helps to satisfy that deeply ingrained desire to be one with nature, not apart from it.

But the appeal of badgers goes far beyond the magic circle of those who love to watch them in the wild. Badgers seem to typify the very essence of the countryside. They are part of our heritage.

Figure 12.7 With regular feeding and patience, wild cubs may be persuaded to feed from the hand

13 The American badger

There is a single species of American badger, *Taxidea taxus*, which is found as far north as southern Canada and ranges well into Mexico in the south. In the west, it reaches the coast from Washington southwards into California and goes as far east as Illinois and Missouri and north-east as far as the Great Lakes. Four sub-species have been recognised by Long (1972):

(1) *Taxidea taxus taxus* from southern, central Canada to the more northerly and central states of America.
(2) *Taxidea taxus jeffersonii* (formerly *neglecta*) from British Columbia down the west coast to California, into Nevada, Utah, Colorado and eastwards to central Montana and Wyoming.
(3) *Taxidea taxus jacksoni* south of the Great Lakes region in Wisconsin, northern Indiana and Ohio.
(4) *Taxidea taxus berlandieri* a smaller sub-species found in the southern states and Mexico.

These sub-species intergrade wherever they meet. There is considerable differences in size, the type species being comparable to *Meles*, but the southern badgers are much smaller, so the head–body length for the species as a whole varies from 420–720 mm and the tail, 100–150 mm. Adult weights (4–12 kg) also vary geographically and also seasonally as in areas which are cold in winter stored fat can account for 30 per cent of body weight. Females are distinctly smaller and lighter.

The colour of the upper parts varies from reddish-brown to greyish, but there is a conspicuous white median stripe which extends from the nose to neck or shoulder, and may continue to the rump in the southern sub-species. Dark patches occur on cheeks and crown. The chin, throat and much of the ventral region are much lighter varying from white to pale reddish buff. The feet are dark brown.

Many of the badger's characteristics are associated with digging. The head is wedge-shaped, the body stocky and low-slung on short but very strong legs, the ears, rounded and protected from soil when digging by stiff bristles and the nictitating membrane is well developed so that it can cover and clean the whole eye surface. The fore feet are partially webbed. This keeps the toes closely together and makes the digging movement stronger

THE AMERICAN BADGER

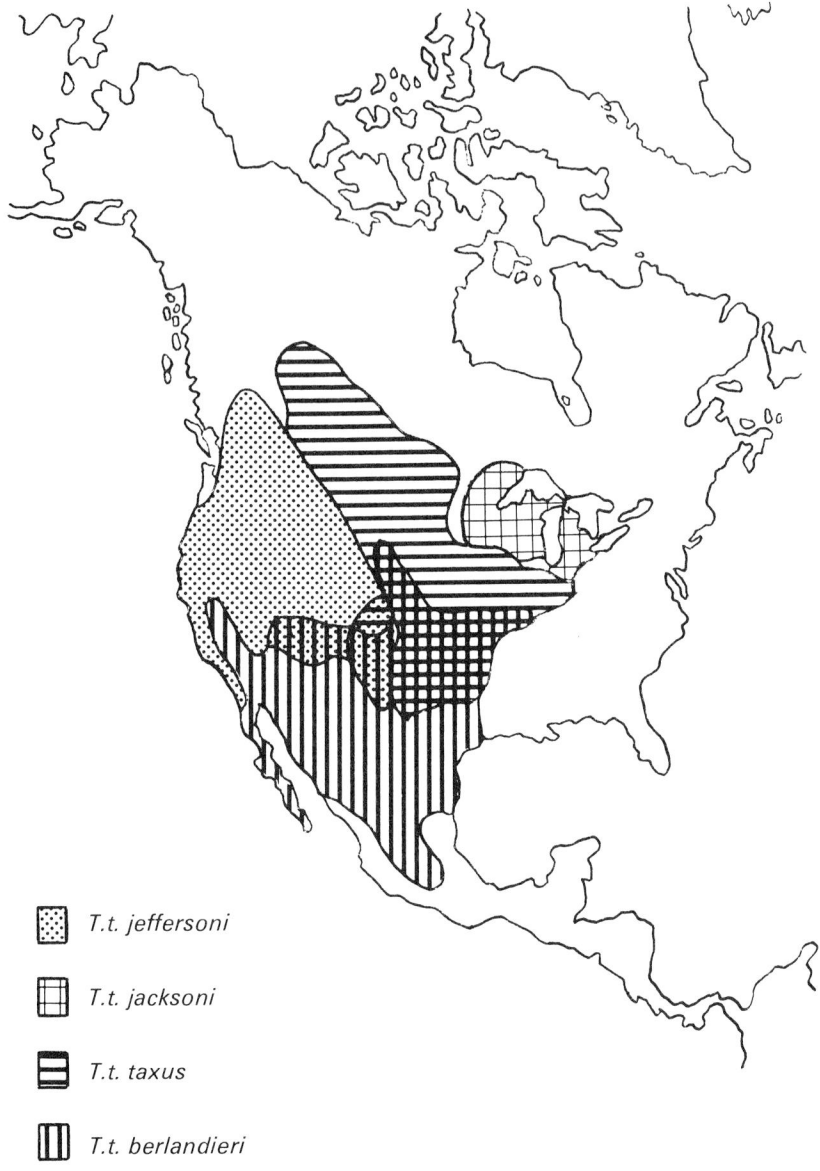

Figure 13.1 Distribution of the four sub-species of the American badger, Taxidea taxus *(After Long and Killingley, 1983)*

- T.t. jeffersoni
- T.t. jacksoni
- T.t. taxus
- T.t. berlandieri

and more effective. The claws of the fore feet are long and powerful: they grow at a faster rate than the smaller ones of the hind feet, a compensation for greater wear.

The skull, 113–141 mm in length, is very robust with an exceptionally broad cranium and a high sagittal crest. The dental formula is:

$$I\frac{3}{3} \, C\frac{1}{1} \, P\frac{3}{3} \, M\frac{1}{2} = 34.$$

Figure 13.2 American badger, Taxidea taxus

This shows a reduction in the number of premolars from four, as found in the more primitive ferret badgers, to three. The Eurasian badger is intermediate having three functional, but often vestigial first premolars in addition.

Although the eyes are small their sight appears to be very much better than in *Meles*, perhaps because they are more diurnal. Senses of hearing and smell are very acute and are of great importance in locating prey in their burrows. They make hisses when confronted and may make pig-like grunts, but their vocalisation appears to be restricted.

The various sub-species inhabit a wide range of habitats, but the most usual are dry open plains and deciduous woodland. They may also be found on farmland and in marshy areas and in the mountainous west may be found as high as 4,260 m in the arctic–alpine zone. In contrast the smaller, southern sub-species may live in hot, inhospitable deserts such as Death Valley and in the arid prairies of Mexico.

In the south they are active throughout the year, but where winters are cold they become semi-dormant for long periods feeding on their stored fat. During this period there may be a drop in body temperature and a slowing of the heart rate. Messick and Hornocker (1981) found that one female emerged only once during a 72-day period of severe cold. They are normally active by night, but in secluded places they are more diurnal than *Meles*. However, in the hot arid regions of their range they are more strictly nocturnal, keeping to their burrows during the day to stay cool and avoid excessive loss of water.

DIGGING

Digging is a major activity. They excavate burrows as shelters and for breeding, but in particular they dig extensively during hunting operations.

They may travel quite widely over their ranges during the summer and may use a new burrow each day. In the autumn they tend to revert to previously used dens and occupy a single one for the winter. Breeding dens tend to be larger than those used as temporary refuges and so have larger spoil heaps outside. They may be 4–5 m long, reach a depth of 2 m and have one or two entrances. They are of simple construction, usually with a single large central chamber containing dry bedding. The main tunnel often bifurcates and joins up again and there may be several short dead-end tunnels and smaller side pockets. Dung is usually deposited in these pockets and is usually covered (Lindzey 1976).

But digging is also a major means of obtaining food. Its favourite prey species are rodents and these are largely taken within their burrow system. The badger often makes numerous trial digs, smelling carefully at each to locate the position of the animal before digging after it. In a good rodent locality a badger may exploit this food source over many weeks and the whole area becomes riddled with excavated holes and extensive heaps of earth.

The speed of digging is phenomenal. When caught in the open a badger can dig itself to safety and disappear within a minute or so leaving a plug of earth to block the burrow behind it.

FOOD AND FEEDING HABITS

Taxidea is largely carnivorous although it will certainly supplement its diet with plant food. In contrast to *Meles* it is primarily a hunter, but above ground it will feed opportunistically on any small invertebrates it comes across. There is often one species of small mammal, usually a rodent, which is a primary prey, the species varying according to habitat. In some places it is ground squirrels, in others, mice, pocket gophers, kangaroo rats, prairie dogs or cottontails. Young skunks, marmots, chipmunks, deer mice, voles and rats are also taken when opportunities occur. Most of these mammals are caught below ground by rapid digging, but occasionally a badger will make a swift dash at prey on the surface or follow for a short distance; cottontails and even hares are caught in this way. Like the Eurasian badger the initial spurt is fast, but thereafter movement is relatively slow.

Insects form a significant part of the diet, particularly in summer. These include crickets, grasshoppers, ground beetles, dung beetles and caterpillars. They will dig for the larvae of cockchafers and ground-nesting bees and have been known to raid wasps' and hornets' nests.

Occasionally they take the eggs of ground-nesting birds, surprise a pheasant when incubating, or take poultry. When there is a surfeit of food the carcasses may be taken back to the burrow or buried and eaten later. Rattlesnakes, garter snakes, lizards, frogs and toads are all taken occasionally, but perhaps more at times when their favourite rodent prey is scarce.

SOCIAL LIFE AND ORGANISATION OF LIVING SPACE

Unlike *Meles* their social life is limited to mating and rearing of young. So the males live solitary lives for most of the year and females and young are usually only together until shortly after weaning. Occasionally there are fights between males over females, but there are no signs of territorial behaviour such as perimeter marking or defence of an area.

This comparative lack of social contact is reflected in the use of the scent glands. There are no sub-caudal glands as in *Meles*, but anal glands are present which produce a pungent and disagreeable odour used in defence. The dung would be to some extent contaminated by anal gland secretion, but as the feces are normally placed in pits and covered with earth it seems likely that they are not important as scent signals.

Messick and Hornocker (1981) found by radio tracking that in good badger country in Idaho the population density could be as much as five adults resident per km^2. Males had home ranges of 2.4 km^2 and females 1.6 km^2. Some male ranges overlapped up to three female ranges and the latter also overlapped. Within these large ranges the badgers lived solitary lives except when breeding. Sargeant and Warner (1972) found that in Minnesota a female used a range of 752 ha in summer and had 50 dens within that area, but in the autumn she used a much smaller adjacent area of 52 ha and in winter this was further reduced to 2 ha.

REPRODUCTION

Mating occurs in summer and autumn and appears to be of long duration although sightings have been few. There is no sign of pair bonding and the sparse evidence available suggests that a male will mate with any females in contiguous ranges. There is a period of delay of about six months before implantation which occurs between December and February—significantly during the quiescent period as in *Meles*. The young are born after about six weeks true gestation in March–early April in a nest of dry bedding in an underground chamber. Litters are usually of two or three (range 1–5). The young at birth are covered in short soft silky hair and are blind and helpless. The eyes open four to six weeks later. Lactation lasts 6–8 weeks possibly longer and the young may disperse soon after being weaned or remain until the approach of the mating season.

Messick and Hornocker (1981) showed that in Idaho, 30 per cent of the young females became mature and were mated during their first year, the remainder as yearlings. Males do not breed until their second year (Wright 1969).

The young are playful and will romp near the entrance of the den while the mother is away hunting. If disturbed they will quickly return to the shelter of the den. Females will defend their young with great ferocity and their bites are strong and capable of inflicting serious damage.

Badgers can live a long time. The record in the wild is 14 years, but in San Diego Zoo a badger lived for 26 years. Greatest mortality occurs during the first two years.

For a more extensive discussion of the American badger the book by Long and Killingley, *The Badgers of the World* (1983), is recommended.

14 Badgers of Asia and Africa

HOG BADGERS—GENUS *ARCTONYX*

There is only one species, *Arctonyx collaris*, which has a wide distribution in south-east Asia. It is found over most of China, Bhutan, Assam, Burma, Thailand, Vietnam and on the island of Sumatra. Thus its distribution is largely tropical and except in parts of China it overlaps very little with *Meles* which has a more northerly range.

Arctonyx resembles *Meles* in many ways, being comparable in size and build except in the northern and Sumatran sub-species which are smaller. The head and body length is 550–700 mm, but the tail is longer, up to 170 mm. Adult weights vary from 7–14 kg. It has the same heavy body as *Meles*, but the head is smaller, and the snout more truncated and mobile with an area of pink, naked skin between the rhinarium and the upper lip. Typically there is a good deal of dark hair on the head including a conspicuous eye stripe and a white stripe runs in the mid-line from the rhinarium to the nape. In some sub-species particularly the one from Sumatra the head is much whiter in the dorsal and anterior regions. The throat and ears are white in all races. The body hair is coarse and in colder regions in winter the underfur becomes thick. The colour is variable, some have a yellowish appearance, others are grey. The fore feet are broad and strong and have long, tough, light-coloured claws.

Hog badgers are found in a wide variety of habitats, but typically in forested areas. They occur in lowland jungles and the more wooded highlands up to at least 3,000 m in northern China. No serious work has yet been done on their behaviour and ecology.

There are a number of sub-species which have been recognised. The most western one, *A. collaris collaris*, occurs in the south-eastern foothills of the Himalayas in Sikkim, Bhutan and into Assam. This is a rather small badger with a head–body length of around 600 mm. It is comparable in size to *A.c. leucolaemus* which is found at even higher altitudes in northern China. *A.c. albogularis* occurs throughout most of China south of the last mentioned sub-species and is rather longer and darker. Another medium-sized badger, *A.c. consul*, occurs in southern Assam and Burma. The largest sub-species is *A.c. dictator* found throughout Thailand, Vietnam and northern Malaya; it is yellowish brown. The Sumatran sub-species, *A.c. hoeveni*, is small and dark, but the top of the head, throat, front feet and tail are conspicuously white.

Hog badgers are shy and largely nocturnal lying up by day in deep burrows of their own making or in natural fissures or under boulders. Their eyesight is poor, but their olfactory sense is extremely well developed and most food appears to be found by scent. They are true omnivores, using their sensitive snouts for rooting in the manner of pigs. They dig up succulent roots and take many kinds of small invertebrates including earthworms. Like *Meles* they appear to be foragers rather than hunters. They are also fond of fruit and in captivity will eat meat, so it seems probable that in the wild they will eat any small animal they can catch.

They are playful, especially when young, and when attacked by a predator they fluff out their guard hairs and display their white throats, defending themselves with teeth and claws and making menacing growls. The skin is very loose and tough, making it very difficult for another animal to get a grip without being bitten in return. Their anal gland secretion is very pungent, but it is not known whether this is used in defence as in the stink badger.

Little is known of their breeding habits except that the young are born in a burrow. There is a record for the sub-species *A.c. leucolaemus* of a litter of four new-born young in April. It was also reported by Parker (1979) that a captive pair from China brought up in Toronto Zoo had two young in February 1977; later, various matings occurred that year and a litter of four was born in February 1978. It is considered likely that delayed implantation occurs in this species and that the true gestation is not more than six weeks.

INDONESIAN STINK BADGER OR TELEDU—GENUS *MYDAUS*

There is only one species *Mydaus javensis* which is restricted in its distribution to Java, Sumatra, Borneo and the North Natuna Islands.

These are small badgers well adapted for burrowing and rooting with a head-plus-body length of up to 510 mm, plus a short tail. The head is small and rather pointed due to the long flexible snout. The body is elongated and supported on very muscular short legs, and the feet are armed with strong claws. The toes of the fore feet are united together as far as the roots of the claws—an adaptation for digging.

The general colour is dark brown, but the crown of the head is white and there may also be white patches along the back, but more often a median white stripe takes the place of these patches and stretches from head to tail.

The teledu is mainly found in the mountains and is nocturnal. During the day it seeks refuge in a short simple burrow of its own making. There is little data about its diet but it is said to feed mainly on invertebrates particularly worms and insects.

This species is notorious for its anal gland secretion which for potency rivals that of the skunk. If molested, it will raise its tail and squirt from these glands a pale greenish fluid which is both nauseating and damaging. There are reports of dogs being asphyxiated by the secretion or even blinded if struck in the eye (Walker 1964). Like the anal gland secretion of civets, it has been used in great dilution as a basis for making perfumes.

PALAWAN OR CALAMIAN STINK BADGER—GENUS *SUILLOTAXUS*

The one species in this genus is *Suillotaxus marchei*. It closely resembles *Mydaus javensis* and is considered by Long and Killingley (1983) to merit only sub-generic rank. It is found on two small islands of the Calamian group north and east of Borneo, Palawan and Busuanga.

It is a small animal with a head-plus-body length of up to 460 mm, plus a very short tail. Ian Grimwood (1976), who as far as I know is the only person to have studied this species in the wild, describes it as being

> of a generally chocolate-brown colour (lighter coloured specimens are not uncommon), with a yellowish cap and streak down the back which fades out at the shoulders, and a dirty white and almost bald muzzle which gives it an anaemic appearance. An enormous, hairless and pale skinned anal region comprises (at least when the animal is alarmed) approximately one-third of its total bulk, and stuck vertically on the extreme end of the anal bulge is a 40 mm stump of a tail, almost devoid of hair and cut square across the top, exactly like the funnel of a toy boat. It moves with a rather ponderous fussy walk, but when alarmed can sustain a steady trot for up to 100 m; even then its speed is no more than that of a man walking.

Grimwood goes on to describe how it is capable of discharging its anal gland secretion with great accuracy. When photographing it, some of this yellowish oil fluid hit the lens of the camera from about 1 m! He describes the smell as pungent but not offensive, and says it passes off after half-an-hour or so.

When one of these animals was first touched, it rolled over and shammed dead, but as Grimwood described 'it spoilt the act by continually swivelling its anal region to point in whatever direction its aggressor moved!' However, Grimwood believes that the secretion is not merely defensive as the tracks it had used when travelling are often permeated with its scent.

These stink badgers were found to be surprisingly common wherever they were found and were occasionally used for food by the local population.

THE FERRET BADGERS—GENUS *MELOGALE* (PREVIOUSLY *HELICTUS*)

These rather small badgers of tropical south-east Asia are more primitive than any other living badgers having four premolars in each jaw like the martens.

There are three species: the short toothed ferret badger *Melogale mosquata*, which ranges from Assam across China, the islands of Taiwan and Hainau and into North Vietnam; the large toothed ferret badger *M. personata*, with a more southerly distribution (but also overlaps the previous species over a wide area) including Assam, Burma, Thailand, Cambodia, Vietnam and Java; Everett's ferret badger, *M. everetti*, which is endemic to Borneo.

They are small badgers with a head-plus-body length of 330–430 mm and a tail of 150–230 mm which can be fluffed up when alarmed. The head is

BADGERS OF ASIA AND AFRICA

Figure 14.1 General range of ferret badgers, South-East Asia

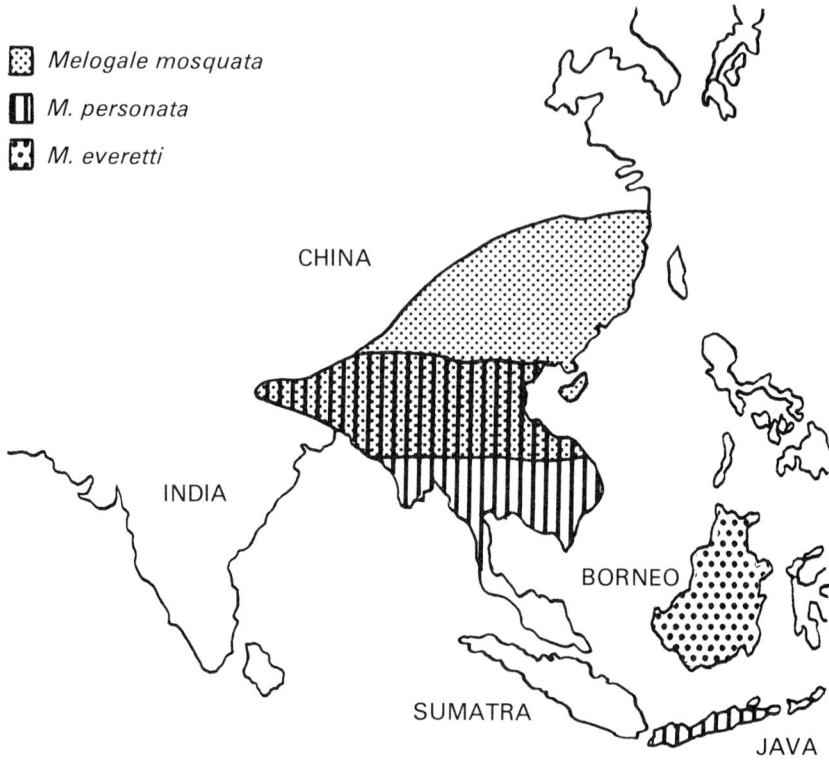

pointed and ends in a pinkish snout and there are various whitish or yellowish markings on head and face including a short but conspicuous transverse band between the eyes which contrasts with a dark patch anterior to it. In addition, there is usually a pale stripe which runs in the mid-dorsal line from the back of the head. The underparts are white or yellowish and the forefeet have strong digging claws.

Figure 14.2 Burmese ferret badger

Ferret badgers live mainly in tropical and sub-tropical forests and wooded hillsides, but they also occur in grassy and more cultivated places. Patricia Marshall (1967) writing about *M. mosquata*, says that their shy and secretive ways aided by acute senses of smell and hearing enable them to live undetected in close proximity to man. She describes them as 'omnivorous, feeding mainly on insects, worms, small birds, young rats and wild fruit. They are nocturnal and spend the day sleeping in a well-hidden burrow or natural shelter, or occasionally on the branches of a tree.' Although they can climb well, it is thought that they spend most of their time foraging on the ground and digging for worms.

Ferret badgers are fearless creatures which may defend themselves fiercely; they emit a pungent odour from the anal glands when attacked.

They have 1–3 young which are born in a burrow in May or June and are suckled for several months.

HONEY BADGER OR RATEL— GENUS *MELLIVORA*

The honey badger is classified in the sub-family *Mustelinae*, partly because it differs from members of the *Melinae* in having only one molar on each side in both jaws. However, it is very badger-like both in form and behaviour. There is one species, *Mellivora capensis*.

It occurs over the greater part of Africa south of the Sahara and its range extends through the Middle East and Arabia as far as India.

Honey badgers are immensely strong, stocky animals about the size of the Eurasian badger and similar to it in build (head-plus-body length 600–770 mm, with a tail of 250–300 mm). They weigh up to 12 kg. They have strong warning colours—a contrast of black and white. All the upper parts are silvery grey or whitish, but the sides and underparts are black or dark brown. When you see them in poor light you notice only the silvery portions so they appear to float in a most uncanny manner. However, in some forest areas in Zaire they may be darker dorsally. The neck and limbs are very muscular and the fore feet are armed with formidable claws. There are no external ears, but the aperture can be closed by muscles which bring the margins together. This happens when digging, keeping the ear free of soil, but may also prevent access of bees or termites when raiding their nests.

The skin is extremely loose and as tough and hard as rubber—an effective protection against most stings, porcupine quills and the bites of other animals. The skin of the throat region is particularly thick—up to 6 mm. Kingdon (1977) considers this to be a special protection during fights with other honey badgers. They can also defend themselves with their extremely strong jaws and by secreting a vile-smelling substance from the anal glands. Consequently, they have no enemies apart from man. But man too has reason to fear the honey badger as there are records of people being viciously attacked and badly bitten. Honey badgers know no fear and will attack even the largest and most ferocious animals when defending their young, making most menacing grunting growls. Cowie related how three honey badgers saw off four half-grown and three sub-adult lions from a wildebeest kill and they have been known to attack sheep and horses. In Kruger National Park an adult buffalo, wildebeest and waterbuck were found dead from loss of

Figure 14.3 Honey badgers

blood after honey badgers had attacked them in the scrotum. They have a considerable reputation as castrators.

Sylvia Sykes (1964) who brought up a honey badger, recounts how, when fighting a dog, once it got a grip it held on regardless of being tossed about, jerking and twisting inside its loose skin until the dog collapsed with exhaustion. It is this tenacity and endurance which makes them such formidable opponents and allows them to overcome large prey.

They feed mainly at night although they may be seen occasionally by day near their dens and when travelling. Characteristically they investigate with their nose any hole or cranny for potential prey. They hunt on their own or in pairs. On one occasion I saw three together of about the same size and there are accounts of larger numbers, but usually these are when a good regular food source attracts them from a wider area. When you see them on their travels they seem to amble along quite slowly, but they can keep up this pace for long periods. Their home range appears to be large.

They are mainly carnivorous but will eat fruits and other vegetable matter. They kill many kinds of mammal such as porcupines, hares, rodents and the young of larger antelopes. Large prey is taken exceptionally. They will also take ground-nesting birds and their eggs, raid crocodile nests and kill snakes, lizards, tortoises and turtles; the latter may be caught under water as honey badgers are good swimmers. They will often eat carrion and may be attracted to safari camps where scraps are discarded. But in spite of all this diversity it is insects that provide a major source of food at certain times of the year; these include dung beetles, wood-boring beetles, ants, termites and the contents of bees' nests; termites and bees are of special importance.

The honey badger has a highly specialised technique for raiding bees' nests. When faced with the fury of a disturbed colony and the probability of a host of stings it uses its anal gland secretion to subdue the bees. The anal glands can be everted through the anus and used to expel a nauseating secretion which acts rather like an anaesthetic, causing some bees to flee and others to become moribund. The badger then tears away the bark to expose the nest and devours the comb containing honey and larvae. Honey badgers are not immune to stings and occasionally they have been found dead within hives having been killed by the bees. Raiding hives can be a major problem in some regions. Kingdon relates that during a single year in Tanzania 2,700 hives out of a total of 24,000 were damaged by honey badgers!

When hunting for bees' nests there is a well authenticated association between the greater honey guide—a bird of the African bush—and the honey badger. When one of these birds finds a bees' nest it is usually unable to get at the larvae and bees' wax on which it feeds, so it may seek out a honey badger. It attracts its attention by repeatedly calling and swooping near, displaying the white on its tail. The badger then follows the bird to the tree which contains the bees' nest. If out of reach, the badger may even climb to get at it. When the nest is exposed the badger takes its fill, but there is always sufficient left for the bird to feed upon.

Honey badgers will also attack termite nests and eat vast numbers of these insects. In some species, the soldier caste are very aggressive and will inflict painful bites and stings. If attacked in this way Kingdon relates how the badger will roll on the ground and dislodge the attackers and at the same time emit much anal gland secretion as a defence.

Anal gland secretion is not only used when attacking the nests of bees and termites. Like all mustelids it has an important function in marking out features within its home range. It has been seen marking tree roots, holes, crevices, tufts of grass and patches of bare ground whilst foraging, particularly near its den. Young raised in captivity regularly set scent on their owners, so in the wild this may happen within families. They often drop their feces into crevices or holes which they may have dug themselves. These may act as scent messages, for as Kingdon remarks, these are just the places other honey badgers would visit.

They can live in very diverse habitats from dense forests to extremely arid scrub although they seem to have a preference for hilly and boulder-strewn country where natural refuges are easier to find. They may also hide up in old termite mounds or aardvark dens particularly during the dry season, using more temporary refuges during the rains.

Mating has been observed in various months of the year and gestation is said to be about six months. If this is correct they must exhibit delayed implantation, a phenomenon usually associated with mammals which have become adapted to a more northerly distribution. As this is not so in the honey badger it is possible that it could be an adaptation to well marked wet and dry seasons. It is said that birth is seasonal in Turkmenia in southern USSR, mating occurring in the autumn with births in the spring. In Africa births appear to be associated with the rains, but the pattern is less clear. They usually have two young (1–4) which are born in a den lined with grassy bedding.

Young born in captivity were blind and naked except for a few facial hairs. One weighed 212 gm. The eyes opened around five weeks and by this time they had the same markings as adults. By two months they could walk normally and by three months were about half size. In the wild they probably come above ground for the first time when 2–3 months old.

In recent years honey badgers in some areas have suffered much persecution because of their habit of attacking bee hives, poultry and other livestock and their numbers have declined markedly. In Tanzania it has been shown that preventive measures, including chemical deterrents can be effective, so it is to be hoped that these may take the place of extermination policies.

BADGERS—PAST AND FUTURE

The fossil record shows that badgers may be traced back to primitive marten-like ancestors and that way back in the Tertiary the ancestors of badgers showed a change in dentition which reflected the evolution from a carnivorous to a more omnivorous diet; shearing teeth became reduced and the molars more flattened. There was also a reduction in the number of teeth from 38 to 34, although the oriental ferret badgers of today still have the more primitive number and *Meles* retains in some specimens vestigial remains of the first premolars.

The *Meles* line appears to have evolved in the temperate forests of Asia spreading west into Europe. Rather primitive forms, but recognisable as badgers, were present some four million years ago in the early Pleistocene period; they may have originated from the Pliocene genus, *Melodon* in China (Kurten 1968). Thoral's badger, *Meles thorali*, was the earliest fossil found in Europe of the genus *Meles*. It was found in France at Saint-Vallier, near Lyons of mid-Villafranchian date—perhaps 2 million years old. Other very similar fossils of the same period were found in China, so this species was probably very widespread. By the early Middle Pleistocene, Europe was inhabited by badgers similar to the modern species; these are now referred to as the sub-species *Meles meles atavus*, Kormos. *Meles* is a common fossil in the Middle and Late Pleistocene, their bones having been found along with those of extinct species such as the cave bear. The earliest record for Britain is a fossil from Barrington in Cambridgeshire estimated at about 250,000 years old.

Petter (1971) considered that the fossil badger, *Arctomeles* from the late Pliocene may have evolved from ancient *Meles* stock and given rise to the modern hog badgers, *Arctonyx*. *Meles* did not spread into the more tropical parts of Asia, possibly because *Arctonyx* filled that niche.

The ancestors of the American badger, *Taxidea*, must have crossed the land bridge between Siberia and Alaska and spread south and west. It is probable that they had their remote origins in forms inhabiting the more tropical parts of Asia which also gave rise to present-day ferret badgers, *Melogale*. An intermediate form found in North America is the fossil *Pliotaxidea* leading to *Taxidea* with its present four distinctive sub-species.

Ferret badgers, *Melogale*, have remained more like the ancestral stock than other badgers, occupying the same tropical habitats. The line of evolution of the honey badger, *Mellivora*, is unknown, but from their teeth it would appear that they are nearer the *Mustelinae* (martens, etc.) than the *Melinae* (true badgers), but sufficiently different from both to be put into a sub-family, the *Mellivorinae*. The remarkable similarities in form, behaviour and ecology are due to parallel evolution due to a comparable mode of life.

To sum up, I cannot do better than quote from Long (1983) who has studied the evidence in depth: 'a picture emerges of long-ago carnivores, slender and quick marten-like climbers, leaving the forests to wander through open woods and over the plains and steppes, digging for shelter, the rearing of young and eventually for food, specialising in fossorial form, generalising in diet and dentition and differentiating into the distinctive genera of today.'

There is still much to be discovered about many of the badger species. Badgers have been a very successful group, exploiting a great variety of habitats on four continents. So far they have held their own with tenacity and have adapted to changes forced upon them by human activities. They have survived man's persecution for many centuries. Today they need our tolerance and understanding if they are to survive along with us. 'It is their world too.' This philosophy of course applies even more to wildlife in general and I am reminded of those words of wisdom from Charles Elton when he defined nature conservation as 'some wise co-existence between man and nature, even if it means a modified type of man and a modified kind of nature'.

Appendix I

Skeletal criteria for determining the age of badgers (Modified from Hancox, 1973)

Age	Teeth	Sagittal crest	Sutures	Epiphyses and symphyses
2–6 weeks	Milk dentition	Absent	All open	All open
6–16	Transition from milk to permanent	Temporal line ridges apparent		
14–16	Permanent dentition complete+ several milk teeth retained			
4–7 months	Adult dentition	20–25 mm gap between temporal line ridges	Closure of interfrontals	Ischio-pubic fusion
8–9		10–19 mm gap between ridges	Fronto-parietal closed	Iliac fusion. Hence two half pelvises. Distal humerus epiphysis fusing
10–12		Closure of gap to form single median ridge... about 2 mm high by 12 months	Closing of squamoso-parietal, nasals and zygoma	
1–2 years	Wear: nil–slight	Ht of ridge 2–9 mm (not smooth) bone yet)		Closing of long bone epiphyses (all closed by 21 months)[1]

APPENDIX I

2–3	Wear: nil–slight	Ht 6–10 mm, sides becoming rugose	Closing of pelvic epiphyses
3–5	Wear: slight–moderate	Ht 7–13 mm, posterior overhang developing	Inter pubic symphysis closed
5–10	Wear: moderate–marked	Ht 8–15 mm	Pelvis one unit
10–15	Wear: very marked	Ht 8–15 mm	

1. R.W. Page (unpub.).

Appendix II

Road underpasses for badgers

Michael Clark has kindly supplied me with the following information which may help others who plan to construct underpasses.

(1) Future rural road schemes in Britain are the province of the County Council Highways Department. Major trunk roads and motorway schemes come under the Department of the Environment Road Construction Units. Route maps may be consulted from these sources.
(2) The earlier the proposed routes are surveyed for setts and paths, the more likely the chance of an underpass being incorporated in the scheme.
(3) Help over locating setts may be obtained through the Mammal Society, the County Naturalists Trusts or the Nature Conservancy Council as well as local farmers and landowners.
(4) The detailed plans showing position of setts and paths should be given to the appropriate authority (see 1) for consideration and a meeting on site arranged between road engineer and conservationist to discuss possible action.
(5) The most inexpensive way to divert badgers is by fencing to an existing culvert, farmstock underpass, gravel or similar structure already incorporated in the plan.

APPENDIX II

Specifications

Concrete pipes. 600 mm diameter.

Culverts. According to circumstances but not less than 1 m in diameter and made of galvanised corrugated steel.

Fencing. 50 m rolls of 1.5 m wide, 100-mesh galvanised wire sheep netting fastened to second rail of post-and-rail tanalised wooden fencing by heavy duty staples. At ground level it is turned outwards and pegged or lightly turfed. The vegetation soon grows through. The amount of netting will vary with local conditions.

Appendix III

Badgers and the law in Britain

Under the Badgers Act (1973) as amended by the Wildlife and Countryside Act (1981) and further amended (1985) badgers are now fully protected. The more important aspects of the Act are summarised below:

It is an offence

(1) To wilfully kill, injure or take any badger or attempt to do so.
(2) To cruelly ill-treat a badger, or use badger tongs, or any fire arm other than a smooth-bore weapon of not less than 20 bore, or a rifle using ammunition having a muzzle energy of not less than 160 foot pounds and a bullet not weighing less than 38 grains.
(3) To dig for a badger.
(4) To possess or have under control a dead badger, or part of, or anything derived from a dead badger: to offer for sale, or have in possession or under control any live badger, or to mark, or attach any ring, tag, or other marking device. Exceptions: (1) mercy killing (b) unavoidable killing as an incidental result of a lawful action (c) temporarily tending an injured badger.

The Amendment (1985) to the Badgers Act (1973) now places the burden of proof on the defendant in cases involving sections 1 and 3 (above) where there is reasonable evidence of an offence. In other words, the defendant will be presumed to have committed the offence unless he can show otherwise.
   This amendment ensures that a person using dogs to kill, injure or take a badger or dig for a badger would not be able to escape conviction simply by claiming that a fox or other animal was the quarry.

APPENDIX III

Licence to take badgers. A licence may be granted subject to compliance with any conditions specified in the licence to take, possess, kill, sell, mark or ring badgers for scientific purposes, for zoos and for the prevention of disease. The licence may be granted by the Nature Conservancy Council for all the above except (a) for disease control (b) for prevention of serious damage to land, crops, poultry or other form of property, where the authority is the Ministry of Agriculture, Fisheries and Food.

Enforcement of penalties. A constable may without warrant stop and search any person found committing an offence under the Act and any vehicle or article which that person may have with him, arrest him if he fails to give full name and address and seize any badger or weapon or article used in the offence. The maximum penalty at present is £1,000 for each badger involved.

Further details of the Act may be obtained from Her Majesty's Stationery Office in London.

Bibliography

Ahnlund, H. (1980) 'Sexual maturity and breeding season of the badger, *Meles meles*, in Sweden', *J. Zool. Lond.*, **190**, 77–95
Andersen, J. (1955) 'The food of the Danish badger, *Meles meles danicus*, with special reference to the summer months', *Dan. Rev. Game Biol.*, **3**, 1–75
Anderson, R.M. and Trewhella, W. (1985) 'Population dynamics of the badger (*Meles meles*) and the epidemiology of bovine tuberculosis (*Micobacterium bovis*)', *Phil. Trans. Roy. Soc.*
Arthur, D.R. (1963) *British Ticks*, Butterworths, London
Aune, V.A. and Myberget, S. (1969) 'The present distribution of the badger (*Meles meles*) in Norway', *Fauna*, **22**, 27–33
Barker, G.M.A. (1969) 'Wytham badger survey . . . diet', unpublished
Bateman, J.A. (1970) 'Supernumary incisors in mustelids', *Mamm. Rev.*, **1**, 81–6
Batten, H.M. (1923) *The Badger Afield and Underground*, Witherby, London
Batty, A. and Cowlin, R.A.D. (1969) 'Notes on some Essex badger mortalities', *Essex Nat.*, **32**, 240–1
Baty, F.W. (1952) Letter in *The Field*, 24 October
Benham, P. (1985) 'The inter-relationships between cattle and badger behaviour and farm husbandry practices and their relevance to the transmission of *Micobacterium bovis*', *MAFF* (in press)
Bere, R. (1970) 'The status and distribution of badgers in Cornwall', (unpublished)
Blakeborough, J.F. and Pease, A.E. (1914) *The Life and Habits of the Badger*, Foxhound, London
Bobrinskii, N.A., Kuzelzov, B.A. and Kuzyakin, A.P. (1944) *Key to the Mammals of the USSR*, State Publishing House Soviet Science, Moscow
Bonnin-Laffargue, M. and Canivenc, R. (1961) 'Étude de l'activité du blaireau Européen (*Meles meles*)', *Mammalia*, **25**, 476–84
Bradbury, K. (1974) 'The badger's diet' in Paget, R.J. and Middleton, A.L.V. *Badgers of Yorkshire and Humberside*, Ebor, York
Brown, C.A.J. (1981) 'Prey abundance of the European badger (*Meles meles*) in north-east Scotland', *Mammalia*, **47**(1), 81–6
Budgett, H.M. (1933) *Hunting by Scent*, Eyre and Spottiswood, London
Burke, N. (1963) *King Todd. The True Story of a Badger*, Putnam, London

—— (1964) 'Some observations on the badger (*Meles meles*)', *Trans. Suffolk Nat. Soc.*, **12**, 437–42
Burness, G. (1970a) *The White Badger*, Harrap, London
—— (1970b) 'Seven year watch on a white badger', *Animals*, Jan., 404–11
Burrows, R. (1968) *Wild Fox*, David and Charles, Newton Abbot
Burton, M. (1957) 'Badgers' warnings', *Ill. Lond. News*, 23 Nov., 900
Butterworth, W.C.J.R. (1905) *Victoria History of the County of Sussex*, Constable, London
Canivenc, R. and Bonnin-Laffargue, M. (1966) 'A study of progestation in the European badger (*Meles meles*)', in Rowlands, I.W. (ed.), 'Comp. Biol. Reprod. Mamm.', *Symp. Zool. Soc.*, **15**, 15–25
—— (1979) 'Delayed implantation is under environmental control in the badger (*Meles meles*)', *Nature*, **278**, 849–50
Cheeseman, C.L. (1979) 'The behaviour of badgers', *Applied An. Ethology*, **5**, 193
Cheeseman, C.L. and Mallinson, P. (1979) 'Radiotracking in the study of tuberculosis in badgers', in Aulaner, C.J. and Macdonald, D. *A Handbook of Biotelemetry and Radiotracking*, Pergamon Press, Oxford
—— (1981) 'Behaviour of badgers (*Meles meles*) infected with bovine tuberculosis', *J. Zool.*, **194**, Pt. 2, 284–9
Cheeseman, C.L., Jones, G.W., Gallagher, J. and Mallinson, P. (1981) 'The population structure density and prevalence of tuberculosis (*M. bovis*) in badgers (*Meles meles*) from four areas of south-west England', *J. App. Biol.*, **18**, 795–804
Clark, M. (1970) 'Conservation of badgers in Hertfordshire', *Trans. Herts. Nat. Hist. Soc.*, **27**(2), 1–9
—— (1981) *Mammal Watching*, Severn House Publishers, London
Clements, E.D. (1974) 'National survey in Sussex', *Sussex Trust for Nat. Cons. Mamm. Rep.* for 1970/1
Cocks, A.H. (1903, 1904) 'The gestation of the badger', *The Zoologist*, **7**, 441–3; **8**, 108–14
Cowland, C.M. (1953) 'Badger v fox', letter in *Country Life*, 19 Nov.
Cowlin, R.A.D. (1972) 'The distribution of the badger in Essex', *Essex Nat.*, **33** (1), 1–8
Cox, N. (1721) *The Gentleman's Recreation*, London
Crossland, J.R. (1934) *Britain's Wonderland of Nature*, Odhams, London
Cuthbert, J.A. (1973) 'Some observations on scavenging of salmon carrion', *Western Naturalist*, **2**, 72–4
Dahl, H. (1954) 'Der Norske Grevling', *Univ. Bergen. Arbon. Nat. Renne*, **16**, 5–55
Dines, A.M. (1981) 'Phantom vandals', *Beecraft*, **11**, 263–5
Drabble, P. (1969) *Badgers at my Window*, Pelham, London
—— (1970) 'Aural evidence of spring mating in badgers (*Meles meles*)', *J. Zool. London*, **162** (4), 547–8
—— (1971) 'The function of mutual grooming in badgers (*Meles meles*), *J. Zool. London*, **164** (2), 260
Dunwell, M.R. and Killingley, A. (1969) 'The distribution of badger setts in relation to the geology of the Chilterns', *J. Zool. London*, **158** (2), 204–8
Edwards, M. (1966) *The Badgers of Punchbowl Farm*, Michael Joseph, London
—— (1971) *The Valley and the Farm*, Michael Joseph, London
Evans, H.T.J. and Thompson, H.V. (1981) 'Bovine tuberculosis in cattle in

Great Britain. Eradication of the disease from cattle and the role of the badger (*Meles meles*) as a source of *M. bovis* for cattle', *Animal Regulation Studies*, **3**, 191–216, Elsevier Scientific Publishing Co.

Ewer, R.F. *The Carnivores*, Weidenfeld & Nicolson, London

Fairley, J.S. (1975) *An Irish Beast Book*, Blackstaff, Belfast

Fargher, S. and Morris, P. (1975) 'An investigation into methods of age determination in the badger, *Meles meles*', thesis, London University

Ferris, C. (1985) unpublished data

Findlay, D.C. (1973) in Clark, R.S. *et. al.*, 'Bovine tuberculosis in badgers', PICL *Triennial Rev.*, 1971–3

Fischer, E. (1931) 'Early stages in the embryology of the badger', *Verb. Anat. Ges. Jena*, **40**, 22–34

Frank, H.R. (1940) 'Die Biologie des Dachs', *Z. Jadgk*, **2**, 1–25

Fullagher, P.J., Rogers, T.H. and Mansfield, D. (1960) 'Supernumary teeth in the badger', *J. Zool.*, **133** (3), 494

Gallagher, J. and Nelson, J. (1979) 'Causes of ill health or natural death in badgers in Gloucestershire', *The Veterinary Record*, **105**, 546–51

Gillam, B. 'The distribution of the badger in Wiltshire', *Arch.*, **62**, 143–53

Göransson, G. (1974) 'Automatisk registrering av grävlingens ak twitet vid grytet', *Fauna och Flora*, **5**, 165–70, Stockholm

Gorman, M.L., Kruuk, H. and Leitch, A. (1984) 'Social functions of the sub-caudal scent glands secretions of the European badger (*Meles meles*)', *J. Zool. Lond.*, **204**, 549–59

Graf, M. and Wandeler, A.I. (1982) 'The reproductive cycle of male badgers (*Meles meles*) in Switzerland', *Revue suisse Zool*, **89**, Pt 4, 1005–8, Geneva

—— (1982) 'Age determination in badgers (*Meles meles*)', *Revue suisse Zool.*, **89**, Pt 4, 1017–23

Grimwood, I. (1976) 'The Palawan stink badger', *J. Fauna Pres. Soc.*, **13** (3), 279

Hager, P.D. (1957) 'Badgers in the south-west Herts/Bucks area of the Chilterns', *Trans. Herts. Nat. Hist. Soc.*, 24, 201–8

Hainard, R. (1961) *Mammifères sauvages d'Europe*, Neuchatel

Hampton, C. (1947) 'The badger's funeral', *Field Sports*, **7**, 24

Hancox, M.K. (1973) 'Studies in the ecology of the Eurasian badger (*Meles meles*)', unpublished

—— (1980) 'Parasites and infectious diseases of the Eurasian badger (*Meles meles*): a review', *Mammal Review*, **10** (4), 151–62

Harris, S. (1982) 'Activity patterns and habitat utilisation of badgers (*Meles meles*) in suburban Bristol: a radio-tracking study', *Symp. Zool. Soc. Lond.*, **49**, 301–23

—— (1984) 'Bristol badgers, an urban success', *Living Countryside*, **137**, 2721–3, Orbis, London

Harting, J.E. (1888) 'The badger, *Meles taxus*', *Zoologist*, **3** (12), 2

Henshaw, R.R. (1952) Letter in *Country Life*, 9 May

Hewer, H.R. and Neal, E.G. (1954) 'Filming badgers at night', *Discovery*, March, 121–4

Howard, R.W. (1951) 'Observations on the sexual behaviour of the badger', unpublished

Howard, R.W. and Bradbury, K. (1979) 'Feeding by regurgitation in the badger', *J. Zool. Lond.*, **188**, Pt 2, 299

Humphries, D.A. (1958) 'Badgers in the Cheltenham area', *School Sc. Rev.*, **139**, 416–25

Hunford, D. (1960) 'Friendly with badgers', *Countryman*, **57** (1), 59–66

Jackson, H.H.T. (1961) 'Mammals of Wisconsin', *Univ. Wisconsin Press*, Madison

Jefferies, D.J. (1968) 'Causes of badger mortality in eastern counties of England', *J. Zool.*, **157**, 429–36

—— (1975) 'Different activity patterns of male and female badgers as shown by road mortality', *J. Zool.*, **177**, 505–6

Jense, G.K. and Linder, R.L. (1970) 'Food habits of badgers in eastern South Dakota', *Proc. SD Acad. Sci.*, **49**, 37–41

Jensen, P.V. (1959) 'Lidt om gravlingen', *Naturens Verden*, **11**, 289–320

Jones, G.W., Neal, C. and Harris, E.A. (1980) 'The helminth parasites of the badger (*Meles meles*) in Cornwall', *Mammal Rev.*, **10** (4), 163–4

King, N. and Barker, G.M.A. (1964) 'Badgers of Old Winchester Hill, Hampshire. Report to Nature Conservancy', unpublished

King, R.J. (1964) 'The badger gate', *Q. J. Forestry*, **58** (4), 505–6

Kingdon, J. (1977) *East African Mammals. An Atlas of Evolution in Africa*, Academic Press, London

Kock, le D. (1965) 'Ringa the honey badger', *Animals*, **7** (12), 333–5

Kruisinga, D. (1965) 'Enige winterwaarnemingen aan Dassen en andere Masterachtigen te St. Michielsgestel', *De Levende Natuur*, **68**, 73–83

Kruuk, H. (1978) 'Spatial organisation and territorial behaviour of the European badger (*Meles meles*)', *J. Zool. Lond.*, **184**, 1–19

Kruuk, H. and Parish, T. (1977) 'Behaviour of badgers', *ITE Env. Res. Council*, 1–16

—— (1981) 'Feeding specialisation of the European badger, *Meles meles* in Scotland', *J. Anim. Ecol.*, **50**, 773–88

—— (1982) 'Factors affecting population density, group size and territory size of the European badger (*Meles meles*)', *J. Zool. Lond.*, **196**, 31–9

Kruuk, H., Parish, T., Brown, C.A.J. and Carera, J. (1979) 'The use of pasture by the European badger (*Meles meles*)', *J. App. Ecol.*, **16**, 453–9

Kruuk, H., Gorman, M. and Parish, T. (1980) 'The use of ^{65}Zn for estimating populations of carnivores', *Oikos*, **34**, 206–8

Kruuk, H., Gorman, M. and Leitch, A. (1984) 'Scent marking with the sub-caudal gland by the European badger (*Meles meles*)', *Anim. Beh.*, **32**, 899–907

Kurten, B. (1968) *Pleisticene Mammals of Europe*, Weidenfeld & Nicolson, London

Lampe, R.P. (1976) 'Aspects of the predatory strategy of the North American badger (*Taxidea taxus*)', *Doct. Diss. Univ. Minnesota Lib.*, Minneapolis

Lancum, F.H. (1954) *Badgers' Year*, Crosby and Lockwood, London

Lawrence, M.J. and Brown, R.W. (1973) *Mammals of Britain, their Tracks, Trails and Signs*, Blandford Press, Poole

Likhachev, G.N. (1956) 'Some ecological traits of the badger of the Tula Abatis broadleaf forest. Studies on mammals in Govt. Reserves', *Min. of Ag. USSR*, Moscow

Lindzey, F.G. (1976) 'Characteristics of the natal den of the badger (*Taxidea taxus*)', *Northwest Science*, **50** (3), 178–80

Little, T.W.A., Naylor, P.F. and Wilesmith, J.W. (1982) 'Laboratory studies of *Micobacterium bovis* infection in badgers and cattle', *Vet. Rec.*, **111**, 550–7

BIBLIOGRAPHY

Lloyd, J.R. (1968) 'Factors affecting the emergence time of badgers (*Meles meles*) in Britain', *J. Zool.*, **155**, 223–7

Long, C.A. (1972) 'Taxonomic revision of the North American badger (*Taxidea taxus*)', *J. Mammal.*, **53**, 725–59

Long, C.A. and Killingley, C.A. (1983) *The Badgers of the World*, Charles C. Thomas, Springfield, Illinois

Macdonald, D. (1985) 'Badgers and bovine tuberculosis—case not proven', *New Scientist*, 25 October

Macdonald, D.W. (1980) 'Patterns of scent marking with urine and faeces amongst carnivore communities', *Symp. Zool. Soc. Lond.*, **45**, 107–39

MacNally, J. (1970) *Highland Deer Forest*, Dent, London

Madge, G. (1982) 'Badgers move to a better hole', *The Countryman*, **87** (4), 87–92

Malins, C.W. (1974) *Bully and the Badger*, Yeatman, London

Marshall, P. (1967) 'The Chinese Ferret badger', *Animals*, **10** (8), 357

Messick, J.P. and Hornocker, M.G. (1981) 'Ecology of the badger in south-western Idaho', *Wildlife Monographs*, **76**, 1–53

Middleton, A.D. (1935) 'The food of the badger, *Meles meles*', *J. Anim. Ecol.*, **4**, 291

Milner, C. (1967) 'Badger damage to upland pasture', *J. Zool.*, **153**, 554–6

MAFF (Ministry of Agriculture, Fisheries and Food) (1984) *A review of bovine tuberculosis in Great Britain*, HMSO, London

Moysey, G.F. (1959) 'Badger activity automatically recorded', *Bull. Mamm. Soc.*, **12**, 19–23

Muirhead, R.A., Gallagher, J. and Burn, K.S. (1974) 'Tuberculosis in wild badgers in Gloucestershire: epidemiology', *Vet. Rev.*, **95** (24), 525–55

Murray, R.R. (1968) 'Concern for the badger', *Animals*, **11** (7), 302–4

Neal, E.G. (1948) *The Badger* (4th edn, 1975), Collins, London

—— (1962) 'The reproductive cycle of the badger', *Bull. Mamm. Soc.*, **10**, 15–17

—— (1964) 'The breeding behaviour of badgers', *Animals*, **5** (11), 302–6

—— (1965) 'Implantation in the badger', *Bull. Mamm. Soc.*, **24**, 6–8

—— (1966) 'Ecology of the badger in Somerset', *Proc. Som. Arch. Nat. Hist. Soc.*, **110**, 17–23

—— (1969) 'Badger nests above ground' in *Countryman Wildlife Book*, David and Charles, Newton Abbot

—— (1972a) 'The National Badger Survey', *Mamm. Rev.*, **2** (2), 55–64

—— (1972b) 'Conservation of badgers' in *Everyman's Nature Reserve*, David and Charles, Newton Abbot

—— (1977) *Badgers*, Blandford Press, Poole, Dorset

—— (1982) 'Badgers make hay' in *The Countryman*, **87** (2), 102–6

Neal, E.G. and Harrison, R. (1958) 'Reproduction in the European badger (*Meles meles*)', *Trans. Zool. Soc. Lond.*, **29** (2), 67–131

Neal, K.R.C. and Avery, R.A. (1956) 'Observations on badgers in the Quantocks', unpublished

Newcombe, M.J. (1982) 'Some observations on the badger in Kent', *Trans. Kent Field Club*, **9** (1), 16–30

Notini, G. (1948) 'Biologiska undersökningar över gravlingen, *Meles meles*', *Svenska Jägareförbundet Meddelande*, **13** (Upsala), 1–256

Novikov, G.A. (1956) 'Carnivorous mammals of the fauna of the USSR' in Key to the fauna of the USSR 62, *Trans. Israel Prog. Sci. Trans.* (1962), Jerusalem

Nowak, R.M. and Paradiso, J.L. (1983) *Walker's Mammals of the World Vol. 2*, John Hopkins, Baltimore
Ognev, S.I. (1935) 'Mammals of the USSR and adjacent countries', *Trans. Israel Prog. Sci. Trans III* (1962), Jerusalem
O'Kelly, J.H. (1969) 'Funeral of a badger' in *The Field*, 31 July
Orchard, J. (1958) 'Wildlife and tame', in *Countryman*, **55** (3), 437–8
Paget, R.J. (1980) 'Dormancy of a badger outside the sett entrance', *J. Zoo. Soc. Lond.*, **192**, 558
Paget, R.J. and Middleton, A.L.V. (1974a) 'Some observations on the sexual activities of badgers in Yorkshire in the months December–April', *J. Zool.*, **173** (2), 256–60
—— (1974b) *Badgers of Yorkshire and Humberside*, Ebor, York
Palmer, N.B. (1959) 'Effect of isolation on behaviour of badgers', *Chelt. Grammar Sc. Biology Report*, **3**
Parker, C. (1979) 'Birth, care and development of Chinese hog badgers (*Arctonyx collaris albogularis*) at Metro Toronto Zoo', *Int. Zoo Year Book*, **19**, 182–5
Pearson, O.P. and Enders, R.K. (1944) 'Distribution of pregnancy in certain Mustelids', *J. Exp. Zool.*, **95**, 21
Pease, A.E. (1898) *The Badger*, Lawrence and Bullen, London
Petter, G. (1971) 'Origine, phylogenie et systematique des blaireaux', *Mammalia*, **35**, 567–97
Pickvance, T.J. and Babb, H.E.E. (1960) 'Badgers and badger setts on Bredon Hill, Worcestershire', *Proc. Birmingham Nat. Hist. Soc.*, **20** (1), 41–7
Pitt, F. (1935) 'The increase in the badger (*Meles meles*) in Great Britain during the period 1900–1934, with special reference to the Wheatland area of Shropshire', *J. Anim. Ecol.*, **4**, 1–6
—— (1941) 'The badger in Britain', *J. Soc. Faun. Pres. Emp.*, **42**, 17–21
Pocock, R.I. (1911) 'Some probable and possible instances of warning characteristics amongst insectivorous and carnivorous mammals', *Ann. Mag. N.H.*, **8**, 750–7
—— (1940) 'Hog badgers (*Arctonyx*) of British India', *J. Bombay N.H. Soc.*, **41** (3), 46
Popescu, A. and Sin, G. (1968) 'Le terrier et la nourriture du blaireau (*Meles meles*) dans les conditions de la steppe de Dobrouja', *Centen. vol. Mus. Nat. Hist. Grigore Antiopa*, **8**, 1003–12
Ransome, R.D. (1958) 'Badger expansion around Cheltenham and factors affecting emergence', *Cheltenham Grammar School Report*, **2**, 5–9
Ratcliffe, E.J. (1974) 'Wildlife consideration for the highway designer', *J. Inst. Mun. Engineer*, **101** (11), 289–94
Roberts, T.V. (1893) in *Trans. Herts. Nat. Hist. Soc.*, **57** (6), 160–1
Russell, C. (1967) 'National badger survey—Shropshire', *Bull. Shropshire Trust for Nature Conservation*, 16–18
Sankey, J.H.P. (1955) 'Observations on the European badger (*Meles meles*)', *South-eastern Naturalist and Antiquary*, **60**, 20–34
Sargeant, A.B. and Warner, D.W. (1972) 'Movements and denning habits of a badger (*Taxidea taxus*)', *J. Mamm.*, **53**, 207–10
Satchell, J.E. (1967) 'Lumbricidae', in Burgess, A. and Raw, F. (eds.), *Soil Biology*, Academic Press, London and New York, 259–322
Schlegel, M. (1933) 'Die lungenwormeuche beim dachs', *Berlin Terarztl. Wschr*, **341**, 344

Scott, D.R. (1960) 'The badger in Essex', *Essex Nat.*, **30** (4), 272–5
Shepherd, S. (1964) *Brocky*, Longmans, London
Skoog, P. (1970) *The Food of the Swedish badger*, Svenska Jagareforb, Vitrevy 7
Soper, E.A. (1955) *When Badgers Wake*, Routledge and Kegan Paul, London
—— (1957) *Wild Encounters*, Routledge and Kegan Paul, London
Southern, H.N. (ed.) (1964) *The Handbook of British Mammals*, Blackwell Sc. Pub., Oxford
Speakman, F.J. (1965) *A Forest by Night*, Bell, London
Stubbe, M. (1970) 'Population biology of the badger (*Meles meles*)', *Trans. Internat. Cong. Game Biologist*, Moscow, 544
—— (1971) 'Die analen markierungsorgane der dachses (*Meles meles*)', *Zool. Gart. N.F. Leipsig*, **40**, 125–35
—— (1973) 'Schutz und hege des dachses (*Meles meles*)', in *Buch der hege, Bd 1: Haarwild* VEB Deutcher Landwirtschaftsverlag, Berlin
—— (1980) 'Biometrie und morphologie des mitteleuropaischen dachses (*Meles meles*)', *Saugetierk. Inform 1* (4), 3–26
Sykes, S.K. (1964) 'The ratel or honey badger', *Afr. Wildlife*, **18**, 29–37
Teagle, W.G. (1969) 'The badger in the London area', *The Lond. Naturalist*, **48**, 48–75
Thompson, G.B. (1961) 'The ectoparasites of the badger (*Meles meles*)', *Ent. Mon. Mag.*, **97**, 156–8
Tinelli, A. and Tinelli, P. (1980) 'Le tane di istrice e di tasso', *Proc. La Reserva Presidenziale di Castelporziane*, Segretariat (Generale Della Presidinza Della Republica), Nr Roma, Italy
Tischler, W. (1965) *Agrarokologie*, Fischer, Jena
Tregarthen, J.C. (1925) *The Life Story of a Badger*, Murray, London
Vesey-Fitzgerald, B. (1942) *A Country Chronicle*, Chapman and Hall, London
Walker, E.P. (1964) *Mammals of the World* (4th edn, 1983), Hopkins, Baltimore
Wandeler, A.I. and Graf, M. (1982) 'The reproductive cycle of female badgers (*Meles meles*) in Switzerland', *Revue suisse Zool*, **89** (4), 1009–16, Geneva
Wiertz, J. (1976) 'De voedsel-ecologie van de das (*Meles meles*) in Nederland', *Rijks Instituut voor Natuurbeheer Report*, 79/9, Leersum, Netherlands
Wijngaarden, A. van and Peppel, J. van de (1964) 'The badger (*Meles meles*) in the Netherlands', *Lutra*, **6**, 1–60
Willan, R.L. (1963) 'Unwelcome squatters and their hosts', *Ill. Lond. News*, 3 December, 867
Winsatt, W.A. (1963) 'Delayed implantation in the Ursidae', in Enders, A.C. (ed.), *Delayed Implantation*, Chicago University Press, Chicago
Wood, J.E. (1958) 'Age structure and productivity of a grey fox population', *J. Mamm.*, **39**, 74–86
World Wildlife Fund (1984) *Wildlife Link Report: badgers, cattle and bovine tuberculosis*, Godalming, Surrey.
Wright, P.L. (1969) 'The reproductive cycle of the male American badger (*Taxidea taxus*)', *J. Reprod. Fert. Suppl.*, **6**, 435–45
Zunker, M. (1954) 'L'importance des renards dans le propagation de la rage en Allemagne', *Bull. Office Internat. Epizooties*, **42**, 83–93

Index

acorns 136
activity patterns 95–107
  diurnal 106–7
  periodicity 104–6
  winter 104–5
adaptive radiation 1
age 185
  determination: from skeletal criteria 224–5; from skull features 34; from tooth wear 36–7
aggression 81–3, 150–1
  reaction to 25, 85–6
albinism 26–8
American badgers 2, 3, 4, 209–13
  description 209–11
  digging 211–12
  distribution 3, 210
  food 212
  reproduction 213
  social life 212–13
  sub-species 209–10
anal glands 142, 145, 213–19 *passim*
anthrax 188
ants 122

baculum 30–1
badger gates 206–7
badger watching 4–13
  by following 9
  by sett 6–8
  choosing vantage points 7
  optical aids 5, 10
  research methods 10–14
Badgers Act *see* legislation
bedding 46–7
  airing 54, 55
  availability 77
  collecting 46–8
  factors affecting collecting 48–53
  in nest chamber 163
  insulation function 53
bees 121
beetles 119–20
binoculars 5
  infra-red 10, 12, 13
  light-intensifying 10
biological clock 103, 104
birds and eggs 113, 127–30
birth 159–63
biting 31–3, 82, 150–1
  wounds 150
blackberries 75
blastocysts 159, 175

bovine tuberculosis 187, 201–6
  badgers as vectors 203
brain 87–8
breeding 159–63
  above ground 163
  cycle 159
  regularity 170–1
  see also reproduction
bulbs 136

camouflage 24–5
cannibalism 126
carrion and refuse 131–2
caterpillars 122
cats 126
cattle
  effect on emergence 100
  interactions with badgers 204–5
  see also bovine tuberculosis
cereals 113, 132–5
classification 4
  American badgers 209
  Eurasian badgers 65–6
  ferret badgers 217
  hog badgers 215
claws 16, 17
clay balls 45–6
climbing 19, 20
cockchafers 120
colour 24–8
  see also pelage
communication *see* signals
corms 136
corpora lutea 176
cover 73–5
  effect on emergence 99
crane flies 122
cubs
  birth 159–62
  carrying 164
  early development 163–5
  first emergence 164–5
  growth rate 165–6
  new born 163–4
  play 153–4
  reaction to attack 25
  vocalisation 141
  warning coloration 25

day nests 54–7
death
  causes 187–94
  disposal of bodies 156–7
  mortality rate 185
  on roads and railways

191–2
  vulnerable periods 185
delayed implantation 159, 175–80
  see also implantation
dentition 35–7
  American badger 210
  ferret badger 217
  eruption of milk teeth 35–6
  evolutionary change 222
diet
  analysis 109–10
  birds and eggs 113, 127–30
  carrion and refuse 131–2
  cereals 113, 132–5
  earthworms 113, 117–19, 196
  effect on sexual maturity 169, 171
  fruits, seeds and storage organs 113, 135–6
  fungi 137
  game birds 127–8, 206
  general conclusions 111
  green food 136
  insects 113, 119–22
  mammals 122–7
  molluscs 131
  poultry 128–30
  reptiles, amphibians and fish 130–1
  variation with geographical location 114
    with season and weather 114–16
    with type of ecosystem 117
    with urban conditions 196
  see also feeding habits
digging 44–6
  factors affecting 48–53
diseases 187–8, 201–6
dispersal of young 170
distances travelled 148
distribution
  in Britain 78, 182–3
  in Europe and Asia 65–6
  in world 3
disturbance 99–100
diurnal activity 106–7
dogs 85–6, 190
drought 115–16, 129
drinking 77, 137
dung
  analysis 109–10

INDEX

beetles 119
pits 61–3; as territorial markers 145, 148–50

ears 16
earth stopping 78
earthworms 113, 117–19, 196
elders 74–5, 135
emergence 95–104
  effect of cover 99
  of disturbance 99–100
  of food availability 101
  of light intensity 97–9
  of sett residents 101–2
  of weather 100
  of cubs for first time 164–5
enemies 24, 25, 187
erythristic badgers 26–7
Eurasian badgers 2, 3, 4–208
  sub-species 65–6
eyes 16, 88, 164
eyesight 88–90, 164

facial stripes 25
farming and badgers 200–1
fat 21, 23–4, 179
feces *see* dung
feet 16, 17
ferret badgers 2, 4, 217–19
  description 217–18
  distribution 218
  habits 219
fighting *see* aggression
fish 130–1, 132
flooding 73
food and feeding habits 109–37
  carrying 83, 124–5, 168
  see also diet
foot prints 18, 19
forestry and badgers 206–7
fossil record 222–3
foxes 80–3
frogs 130
fruit 113, 135–6
funerals 156–7
fungi 137
fur *see* pelage

game birds 127–8, 206
gassing 191, 201–2
genitalia 30
geology, affecting sett choice 66–73
glandular secretions 142, 143–5
  see also anal glands, subcaudal glands
glass 138
grooming 155–6
growth rate 165–6

habitat

altitude 76
bedding availability 77
cover 73–5
  exploitation 114
food availability 76
geological factors 66–73
proximity to water 77
seclusion 77–8
slope 75–6
urban 151–2, 195–7
see also setts
habituation 9
  to cattle 204
  to light 52
  to sound 91
hair *see* pelage
hares 124
hearing 90–2
hedgehogs 123–4
hierarchy 151
hog badgers 2, 3, 4, 215–16
  description 215
  distribution 3, 215
  habits 216
honey badger 2, 3, 4, 219–22
  association with honey guides 221
  description 219
  distribution 3, 219
  food and feeding habits 220–1
  reproduction 221–2
honey guides 221
hormones *see* steroids

implantation 159, 175–9
  factors involved 176–9
  reasons for failure 186
  see also delayed implantation
inherited behaviour 165
insects 113, 119–22
interactions with other animals 79–86, 204

jaw musculature and articulation 33

lactation 166–8, 173
lamb killing 125–6
lamping 191
latrines *see* dung pits
legislation 182, 190, 226–7
lice 189
life expectancy 185
litter size 162
living space 148–53, 212–13
locomotion 17–19
longevity 185

mammals
  in diet 122–7
  interactions with 79–86

mange 190
mating 169, 171–6
melanism 26
mice 122
Micobacterium bovis see bovine tuberculosis
moles 123
molluscs 131
monopolised zones 151
mortality *see* death
moult 23
musking 143–5, 150, 165
Mustelidae 1–3

National Badger Survey 66, 182
nests 162–3
  above ground 163
  day 54–7
  temperature of 163
nettles 74–5
nipples 31
nuts 136

oestrus 173–4
  secondary 176
os penis see baculum
osteomyelitis 188

parasites 188–90
partridges 128
paths 57–61
pattern significance 24–5
pelage 22–8
  fluffing up 25, 140
  hair development 23
  moult 23
  variation in colour 25–8
penis 30
persecution 190–1
pheasants 127–8
play 153–5, 196
poisoning 192–4
population density 181–94
  estimation 184
  factors influencing 184–94
  historical 181–2
poultry killing 128–30
promiscuity 174

rabbits
  in diet 124–5
  interactions with badgers 79
rabies 189
ranges 148
  see also territory
ratel *see* honey badger
rats 123
recolonisation 186
recording devices 104–5
regurgitation 168
repellents 197, 200

237

INDEX

reproduction 159–80
  birth 159–63
  delayed implantation 175–80
  dispersal 170
  early development 164–5
  growth rate 165–6
  lactation 166–8
  mating 171–5
  sexual maturity 169–70
  see also breeding
return times 104–5
rhinarium 15, 16
rolling 154

sagittal crest 32–4
scent
  detection 92–4
  setting *see* musking
  signals 142–5
  trails 144
scratching trees 63–4
screaming 140–1
seclusion 77–8
semi-albino badgers 27
semi-dormancy 178–9
senses 87–94
setts 39–44
  abandonment 77–8
  ancient 43–4
  density 182–4
  distribution 182–3
  habitat selection 66–78
  in archaeological sites 72–3
  nest chambers 162
  origin 44
  plants associated 74–5
  seasonal use 74
  signs of occupation 6
  size 41–3
  spoil heaps 39–40, 72
  tenanted by other animals 79–83
  underground labyrinth 40–2
  ventilation 42
  see also habitat
sex
  differences 29–31, 34
  ratio 185
sexual maturity 169–70, 171
sheep 84
shrews 123
signals 139–45
  scent 142–5
  visual 139–40
  vocal 140–2
  warning 24–5
size 21
  of new born cubs 163–4
skeleton 31–4
skin 22, 219
skull 31–4

sleeping 56
smell, sense of 92–4
snakes 130
snaring 191
social life 139–57
  mutual grooming 156
  organisation of living space 148–53; American badger 212–13
  play 153–5
  signalling system 139–45
  social groups 145–8
spoil heaps 39–40, 72
steroids 176, 179
starvation 187
stink badgers 2, 3, 4, 216–17
  Indonesian 145, 216
  Palawan or Calamian 217
stomach analysis 109
strength 16
subcaudal gland 31, 142–4, 150
swimming 20–1

tail 21
  movements 140, 143, 173
  sex differences 29
teeth *see* dentition
teledu *see* stink badgers
telemetry 10–14, 149
territory 148–53
  marking 145, 150
  monopolised zones 151
  size 148–9, 184
  see also ranges
ticks 189–90
toads 130
traffic deaths 191, 197–200

underpasses for badgers 197–200
  specifications 225–6
urban badgers 77, 151–2, 195–7
urination 62, 143, 202, 203

vaccines 205
vibration detection 91
vibrissae 15
vision *see* eyesight
vocalisation 140–2, 174
voles 122

wasps 79, 120–1
water availability 77
weaning 161, 168
weather
  affecting diet 114–16
  affecting emergence 100
weights 21–2
  growth rate 165–6
  of new born cubs 163–4
wolverine 64